Benjamin Russel Hanby

Chapel gems for Sunday schools

Selected from the Snow bird, Robin, Red bird, Dove and Blue bird

Benjamin Russel Hanby

Chapel gems for Sunday schools
Selected from the Snow bird, Robin, Red bird, Dove and Blue bird

ISBN/EAN: 9783337257408

Printed in Europe, USA, Canada, Australia, Japan

Cover: Foto ©berggeist007 / pixelio.de

More available books at **www.hansebooks.com**

ENLARGED EDITION.

CHAPEL GEMS

FOR

SUNDAY SCHOOLS;

SELECTED FROM

The Snow Bird, Robin, Red Bird, Dove and Blue Bird,

By GEO. F. ROOT and B. R. HANBY;

AND FROM

THE LINNET,

By F. W. ROOT and J. R. MURRAY.

WITH ADDITIONAL PIECES

By D. P. HORTON, of Brooklyn, N. Y.

CHICAGO:

PUBLISHED BY ROOT & CADY, 67 WASHINGTON STREET.

SUGGESTIONS.

The Sunday School should open with singing; for there are always tardy ones, and it is better that singing be interrupted, (if it *must* be so,) than reading the Scriptures, or prayer.

It should open, too, with a SONG, in preference to a HYMN, for the former will bear interruption better than the latter. And this is the difference between the two, as we desire the terms to be applied to the Sunday School music, *proper*, which we present in this book.

A HYMN always addresses *the Deity*, and is the only kind of sacred music that is strictly *Devotional*. A SONG may be *concerning* the Deity, but it is never *to Him*, only *about* Him. It is simply *pious* or *religious*. A Hymn has but one theme. A song may have any theme in the range of Christian Doctrine or Experience. Sometimes a SONG rises almost to the dignity of a Hymn; sometimes the two are combined, as in Dismission Hymn, page 21. A Hymn should always be sung in a *subdued*, *reverent* spirit. There is more latitude to the *style of expression* in a Song.

The Hymn comes in most appropriately just before beginning the regular lesson of the hour. *Songs* come properly immediately after the lesson, and before the books and papers are distributed.

A certain class of songs will not be found in this book, notwithstanding their popularity with some. We refer to those that picture this world as a "howling wilderness," a "gloomy desert," &c., and adding expressions of dissatisfaction with the present life, and a desire to "get rid of it," and "be at rest." When children sing such songs, they sing what, to them, at least, *is not true*; the impropriety of which, (if not something worse,) ought to be manifest to *all*. On the other hand, we have sought for songs as far as we could, that impress the duty of cheerful submission to the daily cross, and the daily trials that are doubtless apportioned to each as a part of his needful discipline.

Believing that our patrons do not care to pay for what they already have in two or three different sets of music books in their libraries, we have inserted none of the "old standard favorites," (though they are very convenient for "filling up,") but present words and music *all new*, *original*, and prepared especially for this book, with but one or two exceptions.

Believing, also, that it is not so much the *number of pieces named in the Index*, that constitutes the desideratum, as the *solid intrinsic merit* of those that are presented, we have refrained from filling occasional pages, with words to be sung to tunes in other parts of the book.

As the page is twenty-five per cent. larger than in other books at the same price, we have felt at liberty to use larger, clearer type. The advantage of this in reading new music and words, or read ing in a dim light—(alas, that the light in so many of our "basements" should be so dim!)—will be manifest.

It is customary for authors to indicate certain pieces in their works which will probably be welcomed as favorites.

We are satisfied that authors are poorly prepared to decide this matter.

All the songs in the book, with many others, have previously appeared in a Juvenile Musical Quarterly * which we publish, and which is intended to furnish every kind of song that is proper or useful for children to sing. The contents of these numbers, as they came out, have been, during the past year, tested by our subscribers in all their variety of uses,—around the Family Altar, in the Social Circle, in the Day School, the Sunday School; at the Picnic, the Concert, the Exhibition, &c.; and we have been apprised, and we will add *surprised* at the preferences indicated.

In the town of ——, the children of one of the public schools had mastered and could sing every page of the first No. (Snow Bird.) They were requested on one occasion to call out, each one simultaneously, the name of his favorite. The verdict was almost unanimous for a song which we had feared, (after it had gone to press,) would prove of little value, if not useless.

We have submitted, therefore, to the decisions of the subscribers referred to, and have selected accordingly. It is safe to say that almost every piece in the book is a favorite with those who have seen it in the "Song Birds." * We will only add on this point that the Infant Class Department has received special attention, and that the music therein will perhaps be found pleasing to "children of a larger growth."

It is a remarkable fact, that while almost every denomination has a greater or less number of Sunday School Exhibition Books, abounding with all kinds of Speeches and Dialogues, from grave to gay—no suitable music has ever been provided for these occasions, nor for Sunday School Musical Concerts. Picnics, &c.

We have made this a new and separate department. As a guarantee of its completeness, we will only say that the leading music of the monster Children's Concert at Crosby's Opera House, in this city—in Leavenworth, Milwaukee, and many other places, during the past winter, was selected entirely from "Our Song Birds," and that from these we have chosen the *best only* for Chapel Gems.

G. F. R.
B. R. H.

SONGS OF WORSHIP.

We praise Thee O God

Sweet is the Work.

G. F. R.

MODERATO.

1. Sweet is the work, O Lord, Thy glo-rious acts to sing, To
2. Sweet, at the dawn-ing light, Thy boundless love to tell; And
3. Sweet, on this day of rest, To join in heart and voice With

praise Thy name, and hear Thy word, And grate-ful off-rings bring.
when approach the shades of night, Still on the theme to dwell.
those who love and serve Thee best, And in Thy name re-joice.

Happy hearts children bring.

G. F. R.

MODERATO

1. Hap - py hearts chil-dren bring, Now to God the of - fer - ing;
2. Thankful hearts chil-dren bring, As a trib - ute to their King;
8. Lov - ing hearts chil-dren bring, An - gels bear the of - fer - ing,

Sing His praise, learn His ways On this best, this best of days.
God is near, Fa - ther hear, And ac - cept our hum - ble prayer.
To the Lamb, bless - ed name An - gels catch the joy - ful strain.

God is love, then let us sing Prais-es to our Sa - vior King.
God is love, and chil-dren raise, Thankful hearts in songs of praise.
God is love, and an - gels join, Our glad cho - rus round the throne.

Lord of All.

G. F. R.

MODERATO

1. Al - might - y God we praise in Thee, The
2. In child - hood's bliss - ful days de - scend, In -
3. To Thee our morn - ing song of praise, To

Lord of All.—CONCLUDED.

ev - er - bless - ed Trin - i - ty: And while be - fore Thy
spire us for our be - ing's end; And when our out - ward
Thee our eve - ning prayer we raise; Thy glo - ry sup - pliant

Throne we fall, We hum - bly own Thee, Lord of all.
na - ture dies, Own us Thy chil - dren in the skies.
we a - dore For - ev - er and for - ev - er - more.

Now to Jesus Christ the glory.

G. F. R.

MODERATO

Now to Je-sus Christ the glo - ry, And do - min-ion shall be given;

He is Al - pha and O - me - ga First and last in earth and heaven.

The Pathway to Heaven

1. We have found the way to Heav - en, And our Sa - vior is the
2. He who calm'd the rag-ing bil - lows Can our in - most foes con
3. Gen-tle Shepherd guard and guide us, That our feet may nev - er

Light, We have found the way to Heav - en, He will
trol, He who calm'd the rag - ing bil - lows On - ly
stray, Gen - tle Shep - herd guard and guide us In the

CHORUS.

guide us day and night. And a - cross the roll - ing
He can save the soul. And a - cross the roll - ing
straight and nar - row way. And a - cross the roll - ing

riv - er, To the Land of the For - ev - er, We will
riv - er, To the Land of the For - ev - er, We will
riv - er, To the Land of the For - ev - er, We will

lose the path-way ncv-er, For our Sa-vior is the Light.
lose the path-way nev-er, For our Sa-vior is the Light.
lose the path-way nev-er, For our Sa-vior is the Light.

May Thy Love.—(Closing.)

1 May Thy love, O God our Sa-viour, In-to all our
2. Thou our Fath-er—we a-dore Thee, Thou the Fath-er

hearts de-scend, May Thy wis-dom lead and guide us,
in the Son; God and Fath-er,—Son and Spi-rit,

And from ev-'ry ill de-fend.
Ho-ly Trin-i-ty in One.

Praise to the Lord.

MODERATO.

1. Lord Je - sus, God and Man, For love of men a Child, The
2. Lord Je - sus, God and Man, In this thy ho - ly day To
3. We pray for child-like hearts, For gen - tle ho - ly love, For
4. We pray for sim - ple faith, For hope that nev - er faints, For

Ver - y God, yet born on earth Of Ma - ry un - de - filed.
Thee for pre - cious gifts of grace Thy ran-somed peo - ple pray.
strength to do Thy will be - low As an - gels do a - bove.
true com-mu - nion ev - er - more With all Thy bless - ed Saints.

5. On friends around us here
   O let Thy blessing fall,
   We pray for grace to love them well,
   But Thee beyond them all,

6. O joy to live for Thee!
   O joy in Thee to die!
   O very joy of joys to see
   Thy face eternally!

Father! I go to Thee!

Miss Victoria Hayden.

MODERATO.

1. Fa - ther! I go to Thee! Fa - ther! Oh! show to me,
2. Deep - ly be - liev - ing in Thy love re - liev - ing sin,
3. Lord! let me rest in Thee! Thine ev - er blest to be;

Mer - cy and love Earth - ly my thoughts have been: Oh! make me
Fa - ther, I come! Oh! take Thy er - ring child, Driv - en by
Wise in my youth; Striv-ing Thy will to do; Aid - ing my

pure with-in! Lead me a - bove!
pas - sion wild, Oh! take me home!
fel - lows, too! Cleansed by Thy truth!

4. Father! I kneel in guilt!
Father! I feel Thou wilt
Ne'er let me fall!
Father I'll e'er be thine!
Oh! let Thy mercies shine,
Blessing us all!

Blest are the pure.

G. F. R.

"Blest are the pure in heart for they shall see God." Matt. viii. 5th.

MODERATO.

1. Blest are the pure in heart, For they shall see our God: The
2. The Lord, who left the heavens Our life and peace to bring, To
3. Lord, we Thy pres-ence seek; May ours this bless - ing be; Give
4. All glo - ry, Lord, to Thee, Whom heaven and earth a - dore, To

se - cret of the Lord is theirs, Their soul is His a - bode.
dwell in low - li - ness with men, Their pat - tern and their King.
us a pure and low - ly heart, A tem - ple meet for Thee.
Fa - ther, Son, and Ho - ly Ghost, Our God for - ev - er more.

Gondola. C. M.

ANDANTINO

1. We come in child - hood's joy - ful - ness, We
We of - fer up, O God! our hearts, In
2. We come not as the might - y come; Not
But as the pure in heart should bend; Seek
3. To Thee thou Lord of life and light, A -
We bend the knee we lift the heart, And

come, as chil - dren, free!)
trust - ing love to Thee. } Well may we bend in
as the proud we bow;)
we thine al - tars now. } "For - bid them not," the
mid the an - gel throng,)
swell the ho - ly song. } How blest tho chil - dren

sol - emn joy, At thy bright courts a - bove;
Sav - ior said; But let them come to me;
of the Lord Who wait a - round His throne,

RITARD

Well may the grate-ful child re-joice, In such a Fa-ther's love.
Oh Sav - ior dear we hear Thy call, We come, we come to Thee.
How sweet to tread the path that leads To yon - der heavenly home,

Jesus loves Thee.

[Inscribed to the Clinton Sabbath Schools.]

W. Irving Hartshorn. 13

MODERATE.

1. Je - sus loves thee, go and pray; He in - vites thee, go to-
day: En - ter at the o - pen gate.....
       En - ter now, no long - er wait.....

En - ter at the o - pen gate,
En - ter now, no long - er wait;

Je - sus loves thee, he is nigh, And will hear thy sup - pliant cry.

2. Jesus loves thee, and wilt thou
Grieve his gentle Spirit now?
Be unmindful of that love
That would draw thy soul above?
Jesus loves thee, seek his face,
Taste the riches of his grace.

3. Jesus loves thee, go and pray,
From that love turn not away;
But accept him, and he'll be
More than brother unto thee:
Jesus loves thee, be his friend,
He will keep thee to the end.

A Band of Scholars.

G. F. R.

MODERATO.

1. We are all a band of schol-ars, All a band of youth-ful
2. We are striv-ing for the king-dom, For the dear and bless-ed
3. We will sing of our Re-deem-er, Of the Lord our great Re-

schol-ars, All a band of ear-nest schol-ars, And we
king-dom, For the ev-er-last-ing king-dom, And we
deem-er, Of the lov-ing blest Re-deem-er, In our

love the Sun-day school; 'Tis a place of hap-py
love the Sun-day school; Hum-ble hearts in meek-ness
hap-py Sun-day school; While we learn the touch-ing

meet-ing, Cheer-ful hearts each oth-er greet-ing, From the
bring-ing, Grate-ful songs of praise we're sing-ing, While our
sto-ry, Learn the sweet and won-drous sto-ry Of His

world's dull care re - treat - ing, In our hap - py, hap - py school.
voi - ces sweet-ly ring - ing In our hap - py, hap - py school.
life, and cross,and glo - ry, In our hap - py, hap - py school.

Work in God's Vineyard.
R. B. H.

ALLEGRETTO.

1. Work in God's vine-yard, Je - sus hath call'd thee,Call'd thee from
2. Faith-ful thy God hath prom - is'd sal - va - tion, Faith-ful thy
3. Mourn-er,bow'd down o'er earth new - ly riv - en, Love lays on
4. Youth, in its ar - dor, man-hood, in glo - ry, In - fan - cy,

dark - ness in - to the light; Break - ing the chain that
load of sor - row He'll bear; Lead - ing the con - trite
thee that chast - en - ing rod; Look thro' thy mist - y
life's path all yet un - trod; Child-hood, with dim - ples,

long hath en-thrall'd thee,Work while the day lasts,and work with thy might.
safe thro' temp-ta - tion, Up to the man-sions He goes to pre-pare.
sor - row to heav - en, Rise, then, and toil in the vine-yard of God.
age, with locks hoar-y, All have a work in the vine-yard of God.

Descend dear Savior.

R. L. Frisbie.

EARNESTLY

1. De-scend dear Sav-ior fill our hearts, With heavenly pleasures full,
2. Shine on us from Thy ho - ly throne, Make this Thy fa - vored hour,
3. Cre - ate a - new our hearts of stone, And make us pure with-in;

And bless us with Thy pres - ence here, In our loved Sun - day School;
Build up Thy Zi - on here be - low; Endowed with wondrous power:
Wash us in thine own pre-cious blood, And take a - way our sin:

Teach us the way of ho - li - ness, The path that leads to God;
Dis - till the dews of heaven-ly grace, Re-fresh our thirst - y land;
Grant, Lord, the les - sons of this day, May not for - got - ten be;

Help us to know and do Thy will, As taught us by Thy word.
For ev ery good and per - fect gift Comes from Thy gra - cious hand.
Oh, lead us heaven-ward, and at last, Take us to dwell with Thee.

Children of Jerusalem.

MODERATO

1. Chil-dren of Je - ru - sa - lem, Place the roy - al di - a - dem
2. Come, let ev - 'ry heart and tongue, Join and swell the grate-ful song:
3. Par - ents, Teachers, old and young, All u - nite and swell the song;

On the Sav-ior's head, and raise Sa - cred an-thems, to His praise.
Sweet - er, high - er, let us sing, Loud Ho - san - nas to our King.
High - er, and yet high - er rise, Let the cho - rus reach the skies.

Hark! while youthful voi - ces sing Loud Ho - san - nas to our King:

Hark! while youthful voi - ces sing Loud Ho - san - nas to our King.

Swiftly Glide the Hours.

O. F. R.

MODERATO.

1. Swift - ly glide the hours a - way, Speed - ing
2. Toil and rest a - like he shares, Bless - es
3. If to - day our lives have been Soil'd by
4. In the dark - ness and the light, Keep us

from us day by day; Leav - ing ev - er,
both our joys and cares, Makes them all His
thought or deed of sin; Lord, from us the
ev - er in Thy sight; And to Thy dear

as they move, To - kens of our Fa - ther's love.
good - ness prove, Makes them to - kens of His love.
guilt re - move, Fa - ther, par - don in Thy love.
home a - bove, Fa - ther, guide us in Thy love.

Confession.

O. F. R.

REVERENTIALLY.

1. List - en, O, list - en, our Fa - ther all ho - ly!
2. Pit - y me now, for, my Fa - ther, no sor - row
3. For thy for - give - ness, the gift I am seek - ing,

Hum - ble and sor - row - ful, own - ing my sin;
Ev - er can be like the pain that I know;
Noth - ing, O, noth - ing I of - fer to Thee !

Hear me con - fess, in my pen - i - tence low - ly,
When I re - mem - ber that, all through to - mor - row,
Thou, to my sin - ful and sad spir - it speak - ing,

How, in my weak - ness, temp - ta - tion came in.
Miss - ing the light of thy love I may go.
Giv - ing for - give - ness, giv'st all things to me.

4. Keep me, my Father, O, keep me from falling!
   I had not sinned, had I felt Thou wert nigh;
   Speak, when the voice of the tempter is calling
   So that temptation before Thee may fly.

5. Thoughts of my sin much more humble shall make me,
   For thy forgiveness I'll love Thee the more:
   So keep me humble until Thou shalt take me
   Where sin and sorrow forever are o'er.

Now to the Lord.

B. R. H.

MODERATO

1. Now to the Lord on high, Ye saints your voi - ces raise; Let
2. Here on this ho - ly day Ye mul - ti - tudes re - pair. And
3. Re - joic - ing, or in grief, Come sit and hear His Word; And
4. His ear is quick to hear, His hand is o - pen wide; Each

lit - tle chil - dren throng His courts, And sing the Sav - ior's praise.
pour your swell - ing souls in song, Or lift the hum - ble prayer.
thro' your smiles, or thro' your tears, Look up and see your Lord.
trust - ing soul shall sure - ly find His ev - 'ry want sup - plied.

Doxology. L. M.

G. F. R.

REVERENTIALLY

To God, the Fa - ther, Spir - it, Son, In soul and mind, and be - ing one; Be

glo - ry, praise and ser - vice given, By all on earth, and all in heaven.

Dismission Hymn.

MODERATO

By permission.

1. All to-geth-er, all to-geth-er, Raise, raise the song.
Ere we sev-er, ere we sev-er Friends, school-mates dear.
2. Thus to-geth-er, would we ev-er, Hand join'd in hand,
When life's les-sons and its la-bors, All, all are o'er.

Sweet the grate-ful strains as-cend-ing, From this glad and hap-py
Join this offer-ing to our Fath-er, For his help and pres-ence
Tread the sa-cred paths of du-ty, On-ward to the Bet-ter
May we with Thy ran-som'd mil-lions, Meet Thee on the Gold-en

CHORUS

throng.
here.
Land.
shore.

{ Thou, thou the Giv-er Of all earth-ly good to men.
{ There, oh Thou Giv-er Of all earth-ly good to men,

Oh may we ev-er Mag-ni-fy Thy worth-y name.
Will we for-ev-er Mag-ni-fy Thy worth-y name.

The Power of Prayer.

G. F. R.

MODERATO.

1. When my soul was distress'd and my spir - it was bow'd, And the
2. When my friends had all left me a - lone to my lot, Then I
3. When bil - lows of sor - row did o - ver me roll, Then I

dark waves of trou - ble ran wild; Then I pray'd to the Lord and He
went to my Sa - vior and Friend; And He soothing-ly spake to my
pray'd for His help from a-bove; And He look'd down upon me and

part - ed the cloud, And He look'd down up - on me and smil'd.
spir - it, "Fear not; I am with thee e'en un - to the end."
fill'd up my soul With e - mo - tions of rap - tu - rous love.

Oh the sun - shine drove dark - ness a - way, And
Oh the sun - shine drove dark - ness a - way, And
Oh the sun - shine drove dark - ness a - way, And

freed my glad heart from its pall; And I wish'd, oh I wish'd that the
freed my glad heart from its pall; And I wish'd, oh I wish'd that the
freed my glad heart from its pall; And I wish'd, oh I wish'd that the

whole world would pray For the smile of the Lord on us all.
whole world would pray For the smile of the Lord on us all.
whole world would pray For the smile of the Lord on us all.

"Savior we Thy Children Gather."

Opening Hymn. MODERATO. B. R. H.

1. Sa-vior we, Thy children gather, In Thy blessed courts to-day,
2. Thou wilt trace the path before us, May we walk and never stray,
3. We would clasp Thy hand for-ev - er, In the darkness as the day,

Seeking Thee our God, our Father, Thee, the Life, the Truth, the Way.
If Thy lov - ing care is o'er us Thou, the Life, the Truth, the Way.
Serv-ing Thee with fixed endeavor, Thee, the Life, the Truth, the Way.

Jesus by the Sea.

REVERENTIALLY.

1. O I love to think of Je-sus as he sat be-side the sea;
2. O I love to think of Je-sus as he walk'd u-pon the sea;
3. O I love to think of Je-sus as he walk'd be-side the sea;

Where the waves were on-ly mur-m'ring on the strand; When he
When the waves were roll-ing fear-ful-ly and grand; How the
Where the fish-ers spread their nets u-pon the shore; How he

sat with-in the boat, on the sil-ver wave a-float
winds and waves were still, at the bid-ding of his will,
bade them fol-low him, and for-sake the paths of sin,

While he taught the wait-ing peo-ple on the land.
While he brought his lov'd dis-ci-ples safe to land.
And to be his true dis-ci-ples ev-er-more.

CHORUS.

O I love to think of Je - sus by the sea;
O I love to think of Je - sus by the sea;
O I love to think of Je - sus by the sea;

O I love to think of Je - sus by the sea, And I
O I love to think of Je - sus by the sea, How he
O I love to think of Je - sus by the sea, And I

love the pre-cious Word, Which he spake to them that heard,
walk'd u - pon the wave, His be - lov - ed ones to save,
long to leave my all, At the dear Re - deem - er's call.

While he taught the wait - ing peo - ple by the sea.
While he brought them safe - ly o'er the storm - y sea.
And his true dis - ci - ple ev - er - more to be.

The Risen Lord.

B. R. H.

1. The fish - ers sat with - in their boat, The long, long, wea - ry night; And
2. A form sub-lime stood on the shore, A - mid the melt - ing gloom; It
3. And O! what won-drous tid-ings then! That Je - sus who was slain, Had

hoped and toiled and watched their nets, 'Till morn-ing's dawn-ing light, And then up - on the
was the form of Him they loved, All glo - rious from the tomb: And then up - on the
burst the might-y bars of death, And conquered life a - gain; And still up-on the

si - lent air, They heard that voice once more: That woke such thrills of bliss and
si - lent air, Rang out those tones once more: That woke such thrills of bliss and
si - lent air, We hear that voice once more; It calls us with love-freighted

CHORUS.
Repeat p p.

love In their wea - ry hearts be - fore. Come chil - dren, toil no long - er,
love In their wea - ry hearts be - fore, Come chil - dren, &c.
words, As it called to them be - fore. Come chil - dren, &c.

Through night's lin-g'ring gloom; For the morn-ing sweet is dawn-ing, O'er the conquered tomb

The morning Light.

G. F. R.

1. Bright shine the rays of the beau - ti - ful morn - ing;
2. See from the East, on the moun - tain tops gleam - ing,
3. Wake from thy slum - ber, brave son of the o - cean!
4. Je - sus, Im - man - uel, we hail Thy ap - pear - ing;

Swift roll the shad-ows of dark night a - way: Hail to the light of mil-
Sun-beams re-splen-dent; o'er val-ley and plain Shine the bright rays, in a
Rouse from thy leth- ar - gy, child of the land! Spring each to du - ty! the
By Thy good Spir - it il-lumine each heart; "Light of the world," with Thy

len - ni - al dawn-ing! Come, thou long promised and glo - ri - ous day!
golden flood stream-ing Eastward and Westward, o'er landscape and main.
world in com-mo-tion Calls thee to la - bor with heart and with hand.
mel low rays cheering, Bid sin and dark-ness for - ev - er de - part.

The Beauteous Day.

G. F. R.

1. We are watch-ing, we are wait-ing, For the bright pro-phet - ic
2. We are watch-ing, we are wait-ing, For the star that brings the
3. We are watch-ing, we are wait-ing, For the beau-teous King of
4. We are watch-ing, we are wait-ing, For the bright pro-phet - io

day: When the shad-ows, wea - ry shad-ows, From the world shall
day: When the night of sin shall van - ish, And the shad - ows
day; For the Chief - est of ten thou-sand, For the Light, the
day; When the shad - ows, wea - ry shad-ows, From the world shall

CHORUS.

roll a - way. We are wait-ing for the morn-ing, When the
melt a - way. We are wait-ing &c.
Truth, the Way. We are wait-ing &c.
roll a - way. We are wait-ing &c.

beau-teous day is dawn-ing, We are wait-ing for the morn-ing,

The Beauteous Day.—Concluded.

For the gold-en spires of day. Lo! He comes! see the

King draw near; Zi-on, shout, the Lord is here.

Lord Thy Word Abideth.

1. Lord, Thy word a-bid-eth, And our foot-steps guid-eth!
2. When our foes are near us, Then Thy Word doth cheer us,
3. When the storms are o'er us, And dark clouds be-fore us,
4. Who can tell the pleas-ure, Who re-count the treas-ure

Who its truth be-liev-eth Light and joy re-ceiv-eth.
Word of con-so-la-tion, Mes-sage of sal-va-tion.
Then its light di-rect-eth, And our way pro-tect-eth.
By Thy Word im-part-ed To the sim-ple heart-ed.

The young child Jesus.

ANDANTINO

1. I love to see in pic-tured rest, The fair and gen-tle child,
2. I love to think how, old-er still, For so the Scrip-ture saith,
3. O Lord my Sa-vior Thou wilt be My guide for-ev-er sure!

Re-pos-ing on the lov-ing breast, Of Ma-ry sweet and mild:
He sub-ject was to Ma-ry's will, In hum-ble Naz-a-reth;
Oh teach me how to grow like Thee, O-be-dient, good and pure;

I love to think how Je-sus grew Thro' child-hood's sun-ny hours,
If to His par-ents He could give O-be-dience meek and good,
I lean on Thee, Thou Ho-ly One; Watch o'er my fee-ble breath,

A hap-py child like me and you, A-mong the birds and flowers.
I'll try in ear-ly youth to live As He hath taught I should.
And make me like the Faith-ful Son, The child of Naz-a-reth.

Victor's Palm.

G. F. R.

MODERATO

1. Heavenly Fath - er, teach the way, Teach Thy lit - tle child to pray;
2. May the sweet and heavenly Dove, Come and fill my heart with love;
3. Fill my heart with heavenly peace, Bid my fret - ful passions cease;
4. May Thy ho - ly an - gels spread Guardian wings a-round my head;

How to shun the ways of sin, How the crown of Life to win;
Ev - 'ry e - vil pas-sion quell, Ev - 'ry thought of sin dis - pel;
Con - quer all my foes with - in, Still the storm - y waves of sin;
May Thy dear and lov - ing Eye Watch my foot-steps from on high;

CHORUS.

'Till I shout the an - gel psalm; 'Till I wave the vic-tor's palm;

Hal - le - lu - jah! Hal - le - lu - jah! Wave the vic-tor's palm.

The Beacon Light.

G. F. R.

ANDANTINO.

1. We are sail - ing o'er an
2. Though the skies are dark a -
3. He will keep it ev - er

o - cean, To a far and for - eign shore; And the
bove us, And the waves are dash - ing high, Let us
burn - ing, From the light - house of His love; And it

waves are dash - ing 'round us, And we hear the break - ers
look to - ward the bea - con, We shall reach it by and
al - ways shines the bright - e t When the skies are dark a -

roar: But we look a - bove the bil - lows, In the
by: 'Tis the light of God's great mer - cy, And He
bove: If we keep our eyes up - on it, And we

dark - ness of the night; And we see the stead - y
holds it up in view, As a guide - star to his
steer our course a - right, We shall reach the har - bor

gleam - ing Of our change - less bea - con light. O, the
chil - dren, As a guide to me and you. O, &c.
safe - ly, By the bless - ed bea - con light. O, &c.

light is flash - ing bright - ly, From a calm and storm - less shore,

Where we hope to cast our an - chor, When our voy - ag - ing is o'er

No Night in the Golden City

H. L. Frisble.

MODERATO

1. There will be no night 'n tne gold - en cit - y Je-
2. There will be no tears in the gold - en cit - y No
8. There will be no death in the gold - en cit. - y. For

ru - sa - lem the fair, On its towers and domes will the
sor - row there for aye, Je - sus cares for all with a
life's broad riv - er springs From th'e - ter - nal throne and life's

Sun in glo - ry, Be ev - er shin - ing there: For the
Fa - ther's pit - y, And wipes all tears a - way: There to
tree in beau - ty, Its leaves of heal - ing bring: And for-

Lord is King, and its light for - ev - er; He, with His presence fills
end - less praise will be turn'd our sad - ness, What un-told joys a - wait,
ev - er - more will its gates of glo - ry, Stand o - pen night and day,

All that heaven - ly coun - try be - yond the riv - er The
When our wea - ry souls to that home in glad - ness, Shall
To re - ceive earth's pil - grims who world a - wea - ry Come

blest e - ter - nal hills. Oh, he waits for us with ten - der pit - y,
pass the crys - tal gate. Oh, he waits &c.
up the shin - ing way. Oh, he waits &c.

In that hap - py land of light and beau - ty. We are

go - ing home to the gold-en cit - y, Je - ru - sa - lem a - bove.

REFRAIN

3

Over the Silent Sea.

B. R. H.

MODERATO

1. There's a bright hap - py home high in heav - en a - bove,
2. There's a glo - ri - fied Form and it stands ev - er-more,
3. Ah! that sweet ho - ly Form is thy Sa - vior, so dear,

'Tis the home of the an - gels the dwell - ing of love,
It will wel - come thee on to the beau - ti - ful shore,
And the smiles of His face light my pil - grim-age here,

Far o - ver the si - lent sea;
Far o - ver the si - lent sea;
Far o - ver the si - lent sea;

sea si - lent sea,

O how sweet to be filled with its glo - ri - ous sight,
There the loved ones of Je - sus are safe in the fold;
Thou canst wan - der through des - erts so wild and so wide:

To sur - vey its broad plains and its pil - lars of light;
And their names are all writ - ten in let - ters of gold;
Thou canst bat - tle the storms and the fierce roll- ing tide,

And to roam thro' its splen - dors with be - ings so bright,
And they drink from the foun - tains of rap - ture un - told,
Thou shalt rest when at last we sit down by His side

Far o - ver the si - lent sea; Far
Far o - ver the si - lent sea; Far
Far o - ver the si - lent sea; Fai

sea si - lent sea,

o - ver the si - lent sea, there's a beau-ti - ful home for thee.
o - ver the si - lent sea, there's a beau-ti - ful home for thee.
o - ver the si - lent sea, there's a beau-ti - ful home for thee.

The Prodigal Son.

G. F. R.

"It was meet that we should make merry and be glad; for this thy brother was dead, and is alive again; and was lost, and is found." LUKE xv. 32.

JOYFULLY.

1. Ring the bells of heav - en! there is joy to - day,
2. Ring the bells of heav - en! there is joy to - day,
3. Ring the bells of heav - en! spread the feast to - day;

For a soul re - turn - ing from the wild; See! the Fa - ther meets him
For the wan-d'rer now is rec - on - ciled, Yes, a soul is res - cued
An - gels swell the glad tri-umph-ant strain; Tell the joy - ful tid - ings

out up - on the way: Wel - com - ing His wea - ry, wan-d'ring
from his sin - ful way, And is born a - new a ran - som'd
bear it far a - way, For a pre - cious soul is born a -

CHORUS.

child. Glo - ry! glo - ry! how the an - gels sing;
child. Glo - ry! &c.
gain. Glo - ry! &c

Glo-ry! glo-ry! how the loud harps ring; 'Tis tho ran-som'd ar - my,

like a might - y sea, Peal-ing forth the an-them of tho free.

I will lift up mine eyes.—CHANT. G. F. R.

1. I will lift up mine eyes unto the hills from whence cometh my help.
2. He will not suffer thy foot to be moved. He that keepeth thee will not slumber.
*3. The Lord is thy keeper, the Lord is thy shade upon thy right.... hand.
4. The Lord shall preserve thee from all evil, He shall pre-serve thy soul.

My help cometh from the Lord which made heaven and earth.
Behold, He that keepeth Israel shall neither slumber nor sleep.
The sun shall not smite thee by day, nor the moon by night.
The Lord shall preserve thy going out and thy coming in from this time forth, and even for ev - er - more.

The Lambs of the Upper Fold.

S. B. H.

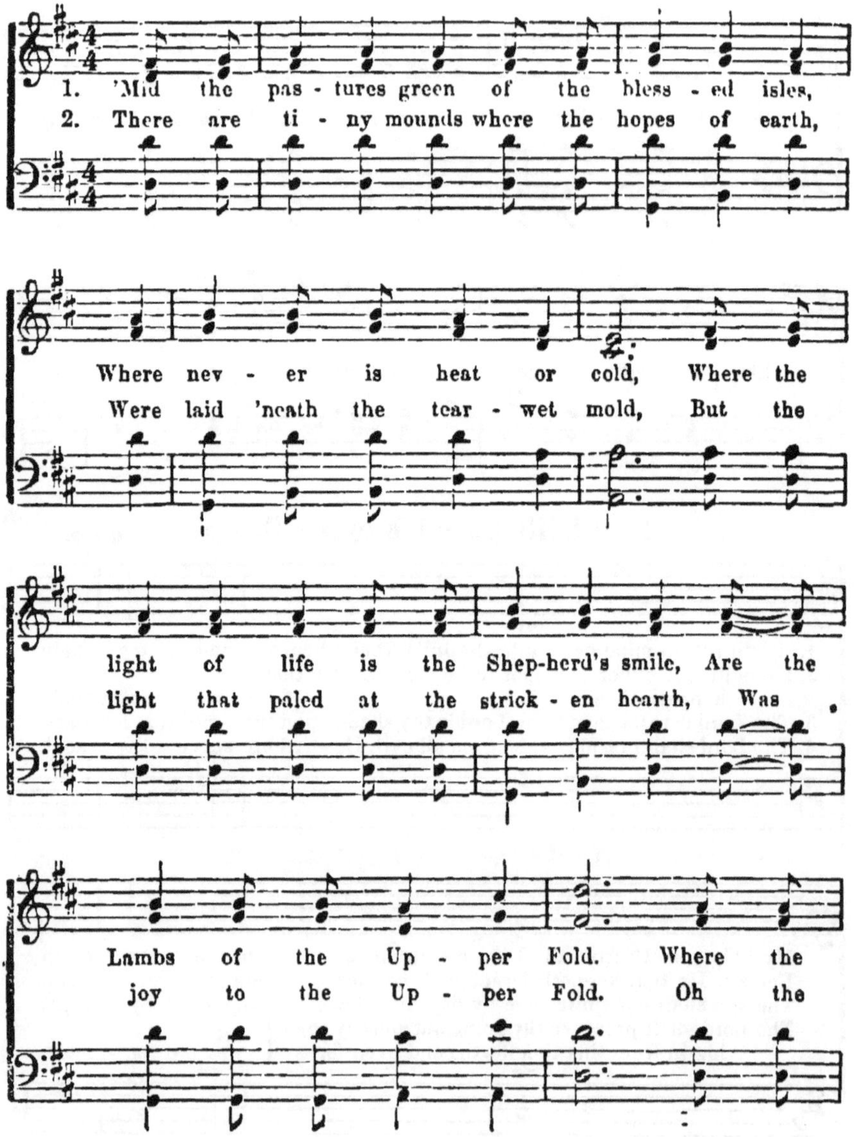

lil - lies blos - som in fade - less spring, And
white stone bear - eth a new name now, That

nev - er a heart grows old, Where the glad new song is the
nev - er on earth was told, And the ten - der Shep-herd doth

song they sing, Are the Lambs of the Up - per Fold. Fold.
guard with care The Lambs of the Up - per Fold. Fold.

Lambs of the Up - per Fold, Lambs of the Up - per Fold,
Lambs of the Up - per Fold, Lambs of the Up - per Fold,

BEAUTIFUL ANGEL

G. F. R.

MODERATO

1. Beau - ti - ful An - gel on
2. Beau - ti - ful An - gel, her
3. Beau - ti - ful An - gel, thrice

pin - ions of light, Wait till I whis - per my Moth - er good night;
sor - row is sore, Weep-ing for one who will weep nev - er - more;
bless - ed art thou! See, there's a smile on the dear pal - lid brow;

List while she calls me her pride and her joy, Folds to her bo - som her
Waft her sweet dreams of the bless-ed a - bove, Tell her, our God is a
To - ken of faith that hath conquered her fears, To - ken that time will have

own lit - tle boy, Hov - er a - round her on pin - ions of light,
Fa - ther of love; On - ly for this, am I stay - ing my flight,
so - lace for tears; Prest to those lips in their ag - o - ny white,

Beantiful Angel.—Concluded.

Moth - er, dear Moth - er, O! kiss me good night.
Moth - er, dear Moth - er, O! kiss me good night.
Moth - er, dear Moth - er, for - ev - er good night.

King.

G. F. R.

MODERATO

1. To Je - sus our God and our King, Our
2. His good - ness is o - ver us all, His
3. Oh then let us all love Him more, And

voi - ces we'll joy - ful - ly raise; And glad - ly His
lov - ing care keeps us from harm, And though we are
try to o - bey His com-mands; That we on the

prais - es we'll sing, Je - ho - vah! the An - cient of Days
ten - der and small, He shel - ters us with His dear arm.
bright shin - ing shore, May join with the pure an - gel bands.

44

Ministering Spirits.

H. W.

ANDANTE.

1. There are un-seen bands of an-gels That are
2. Wouldst thou heed these an-gel whis-pers, To thy
3. There's a spir-it dwells with-in thee, If thou

D. C. Through the day and through the dark-ness, In the

min-is-ters of love, And that bring us sweet e-van-gels,
spir-it day and night? Wouldst thou walk the world sin-dark-ened
heed its yea and nay, That in gen-tle-ness will win thee

heart's un-fath-omed cells, There's a whis-per to our be-ing—

FINE.

From the bless-ed courts a-bove. For they mur-mur words of
With thy raiment pure and white? Wouldst thou know Heav'ns influence
To the high and nar-row way, That will bow thy heart at

There a heaven-ly pres-ence dwells.

warn-ing, Till our e-vil tho'ts de-part, When they
steal-ing, O'er the Ba-bel din of earth? List, the
e-ven, Tho' the scorn-er's lip be curled, That will

D. C.

bring the gold - en morning Of sweet peace with - in the heart
voice-less sweet re - veal-ings Of thy high - er, ho - lier birth.
lead thy heart to Heav - en To the Sa - vior of the world.

The Angel Death. (FUNERAL.)

G. F. R.

ANDANTING.

1. The an - gel Death has borne a - way A *broth - er from our
2. Per - haps our time may be as short, Our days may fly as
3. All need - ful strength is thine to give, To thee our souls ap-

side, Just in the morn - ing of his day, As
fast: O Lord, im - press the sol - emn thought, That
ply; Lord, teach us how we ought to live, And

young as we he died, As young as we he died.
this may be our last, That this may be our last.
make us fit to die, And make us fit to die.

* The words "sister" and "she" may be substituted when appropriate.

My Savior's Voice.

B. R. H.

WITH TENDERNESS.

1. The sky was dark a - bove, The sea was dark be - low; I
2. My heart seem'd dark and wild, A bird that had no rest, A
3. Thus o'er life's path I stray'd, And long'd for rest, sweet rest; But

had no light, no guide, I knew not where to go. A
wea - ry, wan-d'ring child, A sea that found no rest A
found no light, no guide,'Till Je - sus made me blest. A

CHORUS.

voice came down up - on my way, I heard my Sav - ior sweet-ly say,
voice came down up - on my way, I heard my Sav - ior sweet-ly say,
voice came down up - on my way, I heard my Sav - ior sweet-ly say,

RITARD.

"I am thy rest; Come, wea-ry one, come to me, Come and be blest."
"I am thy rest; Come, wea-ry one, come to me, Come and be blest."
"I am thy rest;" I bow'd 'neath my dai-ly cross, And there found rest.

Cast thy Burden on the Lord.—ANTHEM. G. F. R. 47

Cast thy bur - den on the Lord, And he will sus - tain thee;

Cast thy bur-den on the Lord, And he will sus-tain thee, And he will sus-

tain thee, and com-fort thee; He will com-fort thee, He will com-fort

thee: Cast thy bur - den up - on the Lord, Cast thy bur - den up-

on the Lord, and he will sus - tain thee and com - fort thee.

Going Home.

O. F. R.

ANDANTINO.

1. I shall go to Thee my Savior, To my home be-yond the stars,
2. Yet it may be I shall tar - ry Till the noon-tide of life's day,
3. Then I'll go to Thee, my Savior, To my home be-yond the stars,

Where the light is soft - ly gleam-ing Thro' the gates with pearl-y bars;
Or un - til my feet are wea - ry, And my hair is thin and gray;
Where some dear one will be wait-ing By the gate with pearl-y bars;

Where are bands of hap-py child-ren Who have learn'd the ways of truth,
Till the even-ing shadows lengthen, From the glo - ry - tint-ed west,
Where the wel-come lights are gleaming From the mansions of our rest,

And re-member'd their Cre - a - tor, In the days of ear - ly youth.
And the sil - ver chimes shall call me To the morn - ing of the blest.
And the sil - ver chimes are call-ing To the morn - ing of the blest.

Who will Meet me?

1. Who will meet me when I die? Who will lead me to the sky? Who will love me in that land? In that spir-it land, An-gels bright will meet me, An-gels bright, An-gels bright; An-gels bright will meet me, In that spir-it land.

2. When my Savior from on high, Calls my spir-it to the sky, Who will meet me on the strand, Of that spir-it land? An-gels bright, &c.

3. Who will hush my trem-bling heart? Who will heaven-ly joy im-part? Who will love me in that land? In that spir-it land, An-gels bright, &c.

If I were an Angel.

J. F. R.

ALLEGRETTO.

1. If I were an an - gel, with a bright and star - ry
2. If I were an an - gel, with a harp of sweet - est
3. If I were an an - gel, with a crown of pur - est

crown, And a harp that rung the sweet - est notes of
song, I would hast - en on the pin - ions of the
gold, I would come with flow'rs of mer - cy for the

song; I would stoop in love and pit - y, I would
dove; I would guide the err - ing chil - dren from the
poor; I would whis - per of the Sav - ior and the

kind - ly hast - en down, To the wea - ry ones in
paths of sin and wrong, To the folds of Him whose
glo - ries all un - told; I would seek to o - pen

sor - row wait - ing long.   O, sweet mer - cy, sweet an - gel
sweet - est name is Love.   O, sweet, &c.
ev - 'ry pris - on door.   O, sweet, &c.

mer - cy, Love - ly and gen - tle as the dawn, Bend - ing

soft - ly o'er the low - ly, breath - ing hope and heav'n - ly

light, From thy dwell - ing in the ra - diant courts of morn.

O, when shall we be There?

H. L. Frisble

EARNESTLY

1. We are pilgrims seek-ing a cit - y, Where we may safe a - bide;
2. Nev - er ear of mor - tal hath lis-tened To such a glo-rious song;
3. In its streets the an-gels are dwell-ing, And all who here be - low;

Far a-cross the riv - er it li - eth, Be - yond its chill - ing tide;
As that an-them ut-tered for - ev - er, By all that ran-somed throng:
Love the Lord and keep His com-mand-ments, Its bless-ed-ness shall know;

Nev - er hands of mor-tals hath build-ed A cit - y bright and fair,
And the light that fill-eth the cit - y, No mor - tal eye can bear;
He hath formed those heavenly man-sions, For those who faith - ful bear,

As that home we seek as we jour - ney; O, when shall we be there?
For the sun of glo - ry is Je - sus; O, when shall we be there?
Ev - 'ry cross, and crowns will be giv - en; O, when shall we be there?

O, when shall we be There?—CONCLUDED.

REFRAIN

Yes when shall we be there, be there, O, when shall we be there;

Safe at home in the heavenly cit - y, And all its glo - ries share.

Hymn of St. Bernard.

G. F. R.

ANDANTINO.

1. Je - sus the ver - y thought of Thee With sweet-ness fills my breast;
2. Nor voice can sing, nor heart can frame, Nor can the mem -'ry find;
3. O, hope of ev -'ry con-trite heart, O, joy of all the meek;
4. But what to those who find? ah! this Nor tongue nor pen can show;

But sweet- er far Thy face to see, And in Thy pres-ence rest.
A sweet- er sound than Thy blest name, O, Sa - vior of man - kind.
To those who fall, how kind Thou art, How good to those who seek.
The love of Je - sus, what it is, None but His loved ones know.

A Home in Heaven.

MODERATO.

1. We are go - ing home, when the toil - some day In our
2. There are fair - y forms on the oth - er shore Of the
3. We shall all go home when the e - ven comes, As the

Fa - ther's vine-yard end - eth; When the lamps are lit for our
dark and fear - ful riv - er; There are sweet - est notes from the
wea - ry chil - dren gath - er Will the ran-som'd meet in the

home - ward way, As the sun of life de - scend - eth:
"gone be - fore," Whom we'll join no more to sev - er:
gold - en street, In the man - sions of our Fa - ther:

When the bow'rs of green, and the floods be - tween, To the
Should the eye look back, o'er a wea - ry track, From the
They will hear His voice, and their hearts re - joice, That the

dim - ming sight are giv - en; When the King says, "Come," we shall
dawn of life till e - ven, O'er a path of thorns and of
worn earth ties are riv - en; We are go - ing home, and the

all go soon, To the bright, bright home in heav - en.
sun - less moon, Yet the goal we reach is heav - en.
lov'd will come, We are go - ing home to heav - en.

Praise Ye the Lord.

G. F. R.

MODERATO.

1. Praise Him in his hal-low'd dwell-ing, Let thine heart be in - ly swell-ing;
2. Praise Him for the love He bore us, That He trod death's vale be-fore us;
3. Praise Him, rescued brand, who bo't thee, Who hath long and kindly so't thee;
4. That He liv-eth, that He reign-eth, That to hear our pray'r He deign-eth;

While the stars of heav'n are tell - ing, Praise ye the Lord.
That His kind - ly care is o'er us: Praise ye the Lord.
That sal - va - tion He hath bro't thee: Praise ye the Lord.
That the con - trite grace ob - tain - eth: Praise ye the Lord.

Burmah. (MISSIONARY.)

1. A voice that I hear, a-cross the sea, Sings the sweet-est songs
2. Oh! hark to the song that o'er the seas Soft-ly flows a-long
3. For, un-der the palm-trees' love-ly shade, There the dread-ful shrine
4. Oh, chil-dren of God, from east and west, So the hea-then come

of the east to me; It sings of a land where bright suns glow,
on the sum-mer breeze; Oh list-en, and min-gling with its flow,
of the i-dol's made; The land of the east is bright and fair,
to the heaven-ly rest! And Bur-mah be-seech-ing-ly begs to-day,

And the beau-ti-ful blos-soms of Bur-mah blow. Hear it say,
You will hear the sad wail-ing of pain and woe. Hear it say,
But sor-row, and sin and death are there. Hear it say,
That you pit-y and help her and show the way. Hear her say,

Hear it say, "Come to the beau-ti-ful land, a-way!"
Hear it say, "Sit-ting in dark-ness we wait for day!"
Hear it say, "Come, in the night of our need, a-way!"
Hear her say, "Come, ye, and lead us to God we pray!"

Hear it say, hear it say, "Come to the beau-ti-ful land!"
Hear it say, hear it say, "Sit-ting in dark-ness we wait!"
Hear it say, hear it say, "Come in the night of our need!"
Hear her say, hear her say, "Come, ye, and lead us to God!"

Praise the Lord.

FINE.

1. Praise the Lord, who reigns a-bove, And keeps His courts be-low;)
Praise Him for His bound-less love, And all His great-ness show;)

D. c. Him from whom all good pro-ceeds, Let earth and heav'n a-dore.

D. C.

Praise Him for His no-ble deeds, Praise Him for His matchless pow'r;

2. Publish—spread to all around,
The great Immanuel's name:
Let the gospel trumpet sound;
The Prince of Peace proclaim;
Praise Him every tuneful string
All the reach of heavenly art
All the power of music bring—
The music of the heart.

8. Him, in whom they move and live,
Let every creature sing,
Glory to our Savior give,
All homage to our King,
Hallowed be His name beneath,
In the heavens and earth adored;
Praise the Lord in every breath,
Let all things praise the Lord.

58 Suffer Little Children to come unto Me.

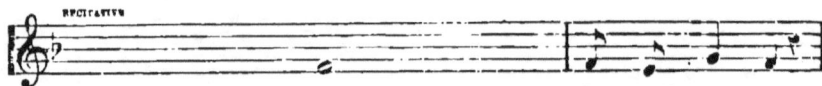

RECITATIVE

And they brought un - to Je - sus young chil - dren that He should touch them

And His dis - ci - ples re - buked those that brought them; But when Jesus saw it He was

ANTHEM

much dis - pleased, And said un - to them; Suf - fer lit - tle

chil - dren to come un - to me, Suf - fer lit - tle chil - dren to

Repeat these two measures on the instrument.

come un - to me, Suf - fer lit - tle chil - dren, Suf - fer lit - tle

chil - dren to come un - to me, and for - bid them not, and for - bid them

not, for - bid them not; for of such is the king-dom of heaven— for of

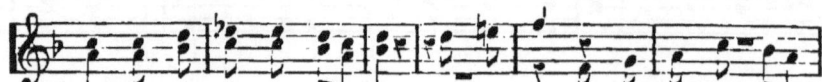

such is the king - dom of heaven, for of such is the king - dom of

heaven, for of such, is the king-dom of heaven, for of such is the

king-dom, the king-dom of heaven, for of such, is the king-dom of

heaven, for of such is the king-dom of heaven, for of such is the

king-dom, the king-dom of heaven, for of such of such is the

king-dom of heaven, for of such of such is the king-dom of heaven.

Hal - le - lu - jah, Hal - le - lu - jah, Hal - le - lu - jah, Praise ye the Lord,

Hal - le - lu - jah, Hal - le - lu - jah, Hal - le - lu - jah, praise ye the Lord,

Hal - le - lu - jah, Hal - le - lu - jah, Hal - le - lu - jah, praise ye the Lord,

Hal - le - lu - jah, Hal - le - lu - jah, Hal - le - lu - jah, praise ye the Lord,,

Hal - le - lu - jah, A - men, Hal - le - lu - jah, A - men, A - men. A - men

"To Him that Overcometh."

J. R. Murray.

ALLEGRETTO

1. "To him that ov - er - com - eth;" O prom - ise of our God.
2. O wea - ry ones, and temp - ted, Rise in the glo - rious might,
3. Then gird ye on the ar - mor; Strive for the vic - tor's palm;

Thou art a glo-rious help - er, A - long our pil - grim road
Of him who died to save you, Arm quick - ly for the fight
Be ear - nest in the con - flict, Stand not in i - dle calm

How can we be dis - couraged, How can we faint and fall,
To him that o - ver - com - eth, God's choic-est gifts are given,
And he that o - ver - com - eth, So saith the Master's word,

Be - hold! our God has prom - ised, We shall in - her - it all.
E - ter - nal life! and glo - ry! The gold - en crown and heaven!
Shall wear the robes of an - gels For - ev - er with the Lord.

The Little Zulu Band.

G. F. R.   61

[Not long ago, a missionary lady in Africa wrote that the little Zulu children were very fond of "Shining Shore," and sang it a great deal (she had translated the words into their language). She also wrote that it would be very pleasant to them to have a song written especially for them, which has been done. This is it.]

1. They come to us from free-dom's land, The gos - pel tid - ings
   And we, a lit - tle Zu - lu band, These tid - ings blest are
D. C. And let us sing it o'er and o'er, Be - fore we reach the

bring - ing; }
sing - ing; } Ah! tell us more of that bright shore, Of
riv - er.

life, and life's great Giv - er;

2. We love to hear the Sabbath chime
   That swells this joyful story ;
   The manger-cradled babe of time,
   Eternal King of glory:
   We love to think of those above,
   In lilied pastures feeding.
   And He, the Shepherd of our love,
   The lambs so gently leading.

3. We look across the darkened wave
   For loved ones, summoned early;
   We see no shadow of the grave,
   Beyond the portals pearly:
   But tell us more of that bright shore,
   Of Him, who will deliver
   From every snare of sin and care,
   And lead us o'er the river.

4. Then we, a little Zulu band,
   Will sing the story olden,
   And pass along the shining strand,
   O'er streets with pavement golden ;
   And o'er and o'er will we adore
   His goodness—life's great Giver,
   And breathe His praise in sweetest
   lays,
   . When we have crossed the river

The Children's Welcome.

MODERATO.

1. "We are com-ing, we are com-ing," 'Twas a soft and sil - v'ry tone,
2. "We are com-ing," 'twas an ech-o Float - ing thro' At - lan - tic's foam,
3. "We are com-ing, we are com-ing, From the moun-tain and the glen!"
4. "We are com-ing, we are com-ing, too, To join the glo-rious band!"

Float - ing thro' the hem-lock for-ests From far Green-land's i-cy zone;
From the chil-dren of the jun - gle, Far off In - dia was their home;
'Twas the chil-dren of New Eng-land,'Twas a glad and heart-y strain;
'Twas a myr - iad voi-ces blend-ing From Pa - ci - fic's gold-en strand;

'Twas the voice of swar-thy lit - tle ones From many a hut of snow,
"We have heard of the An - oint - ed On fair Men-am's vel-vet shore,
Shin - ing ranks of hap-py lit - tle ones Went gai - ly march-ing by,
As the breez - es of the prai-ries Bore the joy-ous notes a-long,

We have heard the won-drous sto - ry, And to Je - sus we would go.
We would turn a - way from i - dols, And the Ho - ly One a - dore.
We have heard the Sav-ior's sum-mons,And will meet them up on high.
Lit - tle chil-dren rush-ing for-ward Swell'd it to a might-y song.

The Children's Welcome.—CONCLUDED. 63

Lyrics:
Hail all hail! Thrice, thrice wel-come, Let the chil-dren come, Come re-ceive a Sav-ior's bless-ing, Come and taste a Sav-ior's love, Come and serve your Lord and Mas-ter, 'Till He wel-come you a-bove, To His heaven-ly home.

2

Festival Hymn.

G. F. R.

M DERATO.

1. Thou who in Thy church of old Sol - emn fes - ti - vals didst place,
2. Is - rael in the Promised Land Bless'd her God that brought her there.
3. Is - rael bless'd the God that sent Sinai's laws of strength and might,
4. Is - rael year-ly brought to mind Mem-ory of her wander-ings drear;
5. At thy fes - ti - vals, O Lord, Thou didst bless the wait-ing hosts,

Smile on us thy lat - er fold; Come our fes - ti - val to grace!
Praise we now the lov - ing Hand That hath made our own so fair.
Thanks we bring that Thou hast lent For our guide, the gospel's light.
Year - ly still our spir - its find No con - tin-uing cit - y here.
Grant us now some quickening word, As of old at Pen - te - cost.

CHORUS

Come, O come to our hearts, Be - lov - ed Lord and King, In -

spire the prai - ses that we give, Ac - cept the thanks we bring.

Thanksgiving.

B. R. H. 65

JOYFULLY

1. Har-vest fields with gold-en glow, La - den branches, bend-ing low:
2. Lord, we know not how to tell All the thanks our hearts that swell;
3. All we have, Oh, Lord! is Thine, Un - to Thee we all re-sign
4. On our gar-ners and our home, Let Thy crowning bless-ing come;

Crowd-ed gar-ners, clos - ing year, Sing, Thanksgiv-ing - time is here.
Hearts that full of grate-ful cheer, Sing, Thanksgiv-ing - time is here.
While Thy chil-dren, Fa - ther dear, Sing Thanksgiv-ing - time is here.
While we, nigh the clos - ing year, Sing, Thanksgiv-ing - time is here.

Father, from whose Hand.

G. F. R.

NOT TO FAST.

1. Fa-ther, from whose hand doth spring Ev-'ry good and per-fect thing,
2. Thou hast placed us here on earth For a high and glo-rious birth;
3. Then, O Fount of ev - 'ry truth, Guard and guide us in our youth;

For the gift of life we raise Songs of grat - i - tude and praise.
And the pre-cious boon hast given To exchange this world for heaven.
Cleanse our souls from ev - 'ry stain, Take them pure to Thee a - gain.

The Shepherds of Bethlehem.

B. R. M.

ANDANTE

1. They were watch-ing on the hill - sides, for the com - ing day,
2. Loud - er swell the joy - ful an - thems from the an - gel throng;
3. Oh, the joy - ful, joy - ful ti - dings! for to you is born,

With the star - ry folds of night a - bove them spread:
O - ver hill and vale the strains en - chant - ed float;
Christ the won - drous Sa - vior and the migh - ty King;

When a glo - ry flashed a - round them, like a ray,
See the won - d'ring shep - herds list - 'ning to the song,
Hail, ye wait - ing na - tions, hail this joy - ous morn!

CHORUS
FASTER, AND WITH ENERGY.

Thro' the pearl-y por - tals on them shed. "Glo - ry to God in the
Trembling, yet re - joic- ing at the sight! "Glo - ry &c.
Hap-py ti-dings now to earth we bring. "Glo - ry &c.

high - est," Come float-ing down the   air; "Glo - ry   to God in   the

high - est!" Seem'd ring-ing ev - 'ry where; "Glo - ry, Glo - ry, Oh,

chil - dren, Come sing that song   a - gain, "Glo - ry to

God in   the high - est! Good - will and peace to   men."

Down from the Skies.

B. R. H.

ALLEGRETTO.

1. Down from the skies bend-ing low o'er the manger, White robed ce -
2. Hail Him ye shep-herds, a-dore Him ye sa-ges, Ho! wait-ing
3. Dark is the path-way be-fore Him and drea-ry, On-ward it

les-tials a-dor-ing-ly throng, Hark! For they her-ald a
Is-rael, still faith-ful though few, Gen-tiles Oh list to the
leads to the cross and the grave, Cheer-ful he treads it though

heav-en-ly stran-ger, Hast-en ye mor-tals to join in their song.
voice of the a-ges, Lo! a De-liv-er-er is com-ing to you.
faint-ing and wea-ry, Thus, on-ly thus, He His lov'd ones can save.

Lit-tle child-ren lisp His grace, Youth-ful voi-ces sound His

4. Weep not Oh stricken ones, when shall enfold Him
   All the deep darkness of Calvary's gloom,
   Soon, soon your tear-blinded eyes shall behold Him
   Walking a God from the gates of the tomb.

CHORUS.—Little children lisp His grace,
   Youthful voices sound His praise,
   Men and angels raise your loud Hosannas to His name,
   Oceans with your fulness roar,
   Earth resound from shore to shore,
   Hallelujah to the Lamb.

MODERATO.

1. With ea-ger step, and hearts in tune, Come, view our Christmas
2. There's gifts for all the pleas-ant throng, Who 'round its won-ders
3. In sum-mer bowers, the sweet Rose Queen Hath fled the win-t'ry

Tree; And feast your eyes, on fai-ry bloom Out-spreading wide, and
stand; The par-ents,and the lit-tle ones, And friends,that swell the
blast; No tree,save yon brave Ev-er-green Stood robed, as on it

free: On twig, and stem, a-loft, a-round, It gleams on ev-'ry
band! Ah! ten-der hands in-spired by love Have gar-land-ed the
passed: So we'll out-ride the storms of sin, If we but choose the

spray; The for-est king is grandly crown'd,To grace our fes-tal day.
tree; And wait-ing ones will grateful prove When shed its fruit shall be.
Right; To heav'nly glo-ries en-ter in Saved by God's love, and might.

Who is He?

MODERATO. The Teacher's part may be uttered in the speech voice.

1. *T.* Who is He in yon - der stall, At whose
2. " Who is He in yon - der cot, Bend - ing
3. " Who is He who stands and weeps At the
4. " Who is He in deep dis - tress, Fast - ing

5. " Lo at mid - night who is He, Prays in
6. " Who is He in Cal - v'ry's throes, Asks for
7. " Who is He that from the grave Comes to
8. " Who is He that on yon Throne, Rules the

CHORUS.

feet the shep - herds fall? 'Tis the Lord, O, won-drous
to His toil - some lot? 'Tis the Lord, &c.
grave where Laz - 'rus sleeps?
the wil - der - ness?

dark Geth - sem - a - ne?
bless - ings on His foes?
heal, and help and save?
world of light a - lone?

sto - ry, 'Tis the Lord, The King of Glo - ry, At His

feet we hum - bly fall, Crown Him, Crown Him, Lord of all.

God is Love.

G. F. R.

1. When light-ly o'er the mountain rill, The twi-light zephyrs move,
2. The bird that trills its evening song, So sweet-ly thro' the grove,
3. The rain-bow in the sum-mer sky Almighty l'ow'r doth prove,

How sweet-ly to the dewy flow'rs They whisper God is Love.
In gen-tle ca-dence seems to say, I'll sing, for God is Love.
Man looks up-on its varied hue, And owns that God is Love.

Little Eyes.

Geo. B. Loomis.

1. Lit-tle eyes, lit-tle eyes. O-pen with the morn-ing light,
2. Lit-tle heart, lit-tle heart Full of laugh-ter, full of glee.
3. Lit-tle hands, lit-tle hands, Bu-sy with the kite or doll.
4. Lit-tle feet, lit-tle feet, Soft your pat-ter, light your load,

Up-ward look, up-ward look, Heav-en's morn is al-ways bright
Beat with love, beat with love For the Lord who bless-es thee.
Learn ye may, work or play, Dai-ly to do good to all.
Do not stray, keep the way, Walk the straight and nar-row road.

Jewels.

"And they shall be mine, saith the Lord of hosts, in that day when I make up my jewels."

MODERATO.

1. When He com-eth, when He com-eth, To make up his jew-els,
2. He will gath-er, He will gath-er The gems for his king-dom;
3. Lit - tle chil-dren, lit - tle chil-dren, Who love their Re-deem - er,

All his jew - els, pre - cious jew - els, His lov'd and his own.
All the pure ones, all the bright ones, His lov'd and his own.
Are the jew - els, pre - cious jew - els, His lov'd and his own.

CHORUS.

Like the stars of the morn-ing, His bright crown a - dorn-ing,

They shall shine in their beau - ty, Bright gems for his crown

We are Little Sunbeams.

G. F. R.

ALLEGRETTO.

1 We are lit - tle sun - beams, Shin - ing and free,
2. We are lit - tle sun - beams, Like those a - bove,
3. We are lit - tle sun - beams, With work to do,

We are lit - tle sun-beams, Hap - py are we; No clouds our
We are lit - tle sun-beams, Warm-ing with love, In - to dark
We are lit - tle sun-beams, May we be true; Where Je - sus

skies o'er cast, No storms are here, Our bright-ness e'er shall last
haunts of woe, Sor - row, and shame, Swift may our bright beams go,
led the way, With foot-steps sure, There we may safe - ly stay,

CHORUS

We will not fear, We are lit - tle sun-beams, Shin - ing and
In Je - sus' name.
There are se - cure.

Little Sunbeams.—CONCLUDED.

free, We are lit - tle sun-beams, Hap - py are we.

Long Ago.

B. R. H.

MODERATO

1, Long a - go, when lit - tle chil-dren Came, the lov - ing Lord to see,
2. Lit - tle children, now to Je - sus Come, with lov-ing, trusting heart;

Je - sus bless'd them, Je - sus loved them, Just such lit - tle
From the spir - it world He sees us, He will bless us

ones as we.
e'er we part.

3. While He on the earth was living,
If He saw one meek and mild,
Gentle, truthful and forgiving,
Well He lov'd that little child.

4. Though He died, He lives in Heaven,
And His care enfolds us still,
To us all His love is given
When we do His Holy will.

Little Sunbeams. Parting Song.

1. Lit - tle sun-beams, we'll a-way From our Sab-bath School to day,
2. When the bells a - gain shall call, May our lit - tle sun-beams, all,
3. When our days on earth are o'er, And we reach the gold-en shore,

Hearts with love are bound-ing free, Hap - pi - er than birds are we.
Here in joy to - 'geth - er meet, Teachers, Scholars all, to greet.
May each lit - tle sunbeam shine, Brighter still, in light di - vine.

CHORUS

Teach-ers dear, a sweet good bye, As we leave you for our homes;

Teach-ers dear, a sweet good bye, Till an - oth - er Sab-bath comes.

Pastures Fair.

Geo. B. Loomis. 81

MODERATO

1. O - ver the hills are the pas-tures fair, And safe the
2. Lead us, and feed us, a hap - py band, There by the
3. O, may He gath - er our wea - ry feet, In - to His

dear lambs are feed - ing there; Come bless - ed Sa - vior, and
hills of the sun - rise land; There by the hills where Thy
pas - tures so fair and sweet; There may we dwell in the

lead our feet, In - to Thy pas-tures so fair and sweet.
loved ones go, Where sweet the wa - ters of life do flow.
gold - en hours, Safe in the bright and e - ter - nal bow'rs.

Come from the Hill-top.

F. R. H.

MODERATO

1. Come from the hill - top, the vale, and the glen, Lights now the
2. Who to the fields or the for - ests would stray Seek-ing their
3. We from the ser - vice of Sin would de - part, Heed-ing Thy
4. Thus when our Sab - baths on earth are no more, We shall be

Sab-bath the land-scape a-gain; Lit-tle feet pat-ter like
pleas-ure at work or at play? Who, when that ban-ner of
man-date of, "Give me thine heart;" Suffer the chil-dren to
with Thee, and love and a-dore: Singing in heav-en that

rain o'er the sod, On in the path to the tem-ple of God.
love is un-furl'd. Turn to the bub-ble-like joys of the world?
"come un-to me." Sa-vior, be-hold at Thy feet here are we.
bright world of bliss Songs that we learn'd on the Sab-baths of this.

On to the tem-ple, On to the tem-ple, On to the tem-ple,

On to the tem-ple, Lit-tle feet pat-ter like

rain on the sod On in the path to the tem-ple of God.

A Helping Savior near.

MODERATO.

1. O, sing to me of that bet - ter land, Of the land of light and
2. Do gloom-y clouds as they dark - ly frown O'er thy steep and toil - some
3. Have friends who 'round thee once fond-ly smil'd Soon-er gain'd the gold-en

bliss; A home all glo - rious a - mid the saints, In a
way; Com - pel a sigh for the fields of light, In a
shore; Do yearn - ings rise in thy sink - ing heart For the

CHORUS.

bright-er world than this. Toil on! toil on! for the Mas - ter needs thee
world of end - less day. Toil on! &c.
lov'd ones gone be - fore. Toil on! &c.

here; Haste, O, haste to the vine - yard wide: Ho - ly an - gels are by thy side, And a help-ing Sav - ior near, A help-ing Sav-ior near.

The New Dress.

ALLEGRETTO.

B. R. H.

1. I missed dear lit - tle Ma - bel from her class and school one
2. Well if sweet lit - tle Ma - bel's moth - er is so ver - y
3. "I will," cried man - y lit - tle ones "and I," cried man - y
4. The plan at once de - cid - ed, then to stud - y as be-

day, And asked the oth - er chil - dren "was she
poor, How ma - ny of these chil - dren now will
more, "We'll soft - ly go some eve - ning dark and
fore, The dress was quick - ly pur-chased and was

sick or gone a - way, "When a wee one blushed and
give a dime or more, To buy an - oth - er
hang it on her door, She will nev - er know who
hung on Ma - bel's door, And eve - ry eye was

stam - mered, "nei - ther sick nor an - y where,
school dress so that Ma - bel shall not stay,
hung it but will rath - er think I guess
bright with joy, and ev - ery heart was gay,

.On - ly her Moth - er keeps her home, she
Need - ing the prop - er clo - thing from her
God in His pi - ty for His child has
When lit - tle Ma - bel smil - ing came in

has no dress to wear," On - ly her Moth - er
school an - oth - er day? Need - ing the prop - er
given her this new dress." God in His pit - y
her new dress next day. When lit - tle Ma - bel

keeps her home, she has no dress to wear.
cloth - ing from her school an - oth - er day.
for his child, has given her this new dress.
smil - ing came, in her new dress next day:

Will the teachers of the infant classes, (the most important of the Sabbath school,) indulge us in a few suggestions from time to time? We have appropriated a portion of this little book to Infant Class Music, as you see; but to make it still more efficient, we desire occasionally to submit some practical directions with regard to its use, and even to offer some suggestions as to other exercises beside those of singing.

We begin by asking whether the repetition of Scripture verses by individual pupils in different parts of the room, (a practice in very general use,) might not be supplanted by something more generally interesting and profitable? To put the question in more general terms, should *any* exercise be allowed to occupy much of the time, that does not require that *all in the room* take part? If a speaker, for instance, undertakes to address the class, he will find it very difficult to hold their undivided attention even for five minutes, without interspersing his remarks with questions to which all are invited to reply.

When the hymn is sung, all the little voices peal out. When the prayer, "Our Father," is uttered, all hands are clasped, and all lips reverently murmur the beautiful petition. Thus far all is delightful; but when the question comes, "Now who has a verse to repeat?" the signal is given for confusion. A score or two of hands go up, and each is anxious to have his turn. The words are uttered almost inaudibly, and so imperfectly as to require frequent correction. The process is a severe tax upon the patience of the majority, and terminating, as it often must, before all have had an opportunity to recite their verses, ends in disappointing some of them. Only the rarest tact and ingenuity on the part of a teacher, will carry a school through such an exercise in an orderly manner. We would, as a *general rule*, (of course subject to occasional exceptions,) say, never introduce an exercise which shall require the little ones very long to keep silent, but insist, when silence is necessary, upon its strict observance. Never introduce exercises except those in which all may take a part, and then see that all engage in it. Thus only, we think, can universal attention be secured, and good imparted to all.

We propose to keep this principle always in view in the preparation of matter for this department, and believing that many would be glad to see this illustrated in all the other exercises beside singing, we venture to offer the following "Infant Ritual," if you please to call it so, inviting the teacher to test its merits by actual experiment.

[The assistant teacher takes her place in the midst of the pupils to teach them the responses, and unite with them in uttering them.]

(Bell taps, and all become quiet.)

PRINCIPAL TEACHER.   Children, listen, for I shall have a question to ask you about what I am reading. (Reads.)
Ps. cxlvii. Praise the Lord.
What does the psalmist tell you to do?
CHILDREN. To praise the Lord.
T. (Reads.)
Praise ye the Lord from the heavens: praise him in the heights. Praise ye him all his angels: praise ye him all his hosts.
T. What are all these to do?
C. Praise the Lord.
Praise him sun and moon: praise him all ye stars of light. Praise him ye heaven of heavens, and ye waters that be above the heavens.
T. And what are these to do?
C. Praise the Lord.
T. Yes, that is right, for the Psalmist adds:
them praise the name of the Lord, for he commanded, and they were created. He hath decree which shall not pass.

Now listen again.

Praise ye the Lord from the earth, ye dragons and all deeps, fire and hail, snow and vapor, stormy wind fulfilling his word. Mountains and all hills, fruitful trees and all cedars, beasts and all cattle, creeping things and flying fowl.

T. What are all these to do?

C. Praise the Lord.

T. Can such things praise the Lord?

C. All his works praise him.

Kings of the earth and all people, princes and all judges of the earth. Both young men and maidens, old men and children. Let them praise the name of the Lord; for his name alone is excellent, his glory is above the earth and heavens.

T. Can children praise the Lord?

C. Yes.

T. When?

C. Always.

T. Where?

C. Everywhere.

T. How?

C. By doing that which is pleasing in his sight.

T. How can you please Him at home?

C. (The assistant teacher will here repeat the following to the children, they repeating after her.) By obedience to our parents....respect for our superiors ....and love for our brothers and sisters.

T. Can you do this in the right spirit, of yourselves?

C. No.....It is God who worketh in us....to will and to do....of his good pleasure.

T. Will God always help us in this way?

C. Yes....if we look to Him and ask Him.

T. How can you please Him at school?

C. By learning our lessons....by loving our teachers....and obeying the rules.

T. How on the play ground?

C. By gentle actions....and kind words.

T. And in doing all these things you would praise the Lord, would you?

C. Yes.

T. How would you praise Him at church?

C. By talking about Him....and by singing and praying to Him.

T. Very well. Let us now do each of these. We will begin by talking in concert about him. We will all say the same words. Let us do so with deep humility, and in a reverent tone.

All recite after me.

I will speak of the glorious honor of thy majesty......and of thy wondrous works......The Lord is gracious and full of compassion;......slow to anger and of great mercy......The Lord is good to all, ...and his tender mercies. ....are over all his works......All thy works......shall praise thee, O Lord......and thy saints......shall bless thee......They shall speak......of the glory of thy kingdom...... and talk of thy power......To make known to the sons of men......his mighty acts......and the glorious majesty......of his kingdom......Thy kingdom......is an everlasting kingdom......thy dominion is throughout all generations......The Lord is righteous......in all his ways......and holy in all his works ......The Lord preserveth......all them that love him......but the wicked......will he destroy.

T. Now we will sing a hymn of praise. I will sing part and you may sing part. Your part will be very easily learned. I will sing it for you. (Teacher sings.)

For his mercy endureth forever ; For his mercy endureth forever.

When the pupils have done this, which they will do almost instantly, the teachers will say :

T. Now you must sing only half of your song at a time. I will sing a few words alone, and then you may put in the first half, then I will sing some more, and you may sing the second half of your song. You see it comes in very prettily.

T. O give thanks unto the Lord ; for he
    is good,   C. For his mercy endureth forever.

T. Oh give thanks unto the God of Gods ;	C. For his mercy &c.
T. Oh give thanks unto the Lord of Lords ;	C. For his mercy &c.
T. To him who alone doeth great wonders ;	C. For his mercy &c.
T. To him that by wisdom made the heavens ;	C. For his mercy &c.
T. To him that stretched out the earth above the waters ;	C. For his mercy &c.
T. Who remembered us in our low estate ;	C. For his mercy &c.
T. And hath redeemed us from our enemies ;	C. For his mercy &c.
T. Who giveth food to all flesh ;	C. For his mercy &c.
T. Oh give thanks unto the God of Heaven ;	C. For his mercy &c.

[NOTE.—If the teacher is not accustomed to chanting, we would say, pronounce the words in the above exercise exactly as if you were talking, only in a monotone, at the pitch indicated, the children beginning the response a third below.]

T. Now, we have talked of the Lord and sung His praises, let us pray to Him.

[Here all kneel or rise to their feet, close their eyes, clasp their hands, and reverently repeat the ʳd's Prayer.]

ʳur Father, who art in Heaven, &c.

O, Sing unto the Lord.

TEACHER.

1. O sing unto the Lord a new song; sing unto the Lord | all | the | earth:
2. Sing unto the Lord, bless his name; show forth his salvation from | day | to | day:

RESPONSE.

Praise ye the Lord.
Praise ye the Lord.

TEACHER.

3. Declare his glories among the heathen, his wonders a-

RESPONSE.

mong all | people: Praise ye the Lord in his ho-ly tem-ple.

4. For the Lord is great, and greatly to be praised; he is to
be feared a- | bove all | gods: } Praise ye the Lord.

5. For all the gods of the nations are idols; but the | Lord
made the | heavens: } Praise ye the Lord.

6. Honor and majesty are before him; strength and beauty
are | in his | sanctuary: } Praise ye the Lord in his holy temple.

7. Give unto the Lord, O ye kindreds of the people, give
unto the Lord | glory and | strength: } Praise ye the Lord.

8. Give unto the Lord the glory due unto his name; bring
an offering, and come in- | to his | courts: } Praise ye the Lord.

9. O, worship the Lord in the beauty of holiness; fear be-
fore him | all | the | earth: } Praise ye the Lord in his holy temple.

10. Let the heavens rejoice, and let the | earth be | glad: Praise ye the Lord.

11. Let the sea roar, and the | fulness there- | of: Praise ye the Lord.

12. Let the field be joyful, and all that | is there- | in: Praise ye the Lord in his holy temple.

13. Then shall all the trees of the wood rejoice be- | fore
the | Lord: } Praise ye the Lord.

14. For he cometh, for he cometh to | judge the | earth: Praise ye the Lord.

15. He shall judge the world with righteousness, and the
| people with his | truth ·   } Praise ye the Lord in his holy temple.

Consider the Lilies.

G. F. R.

MODERATO

1. Con - sid - er how the lil - ies grow, They la - bor not, nor spin:
2. The float-ing cloud, the deep blue sky, The glo-rious morn, the day;
3. His love is Love Di - vine, and far Ex-ceeds our high - est tho't;

Not proud - est kings of earth, we know Such gorgeous ves-tures win:
The fall - ing leaf, the zephyr's sigh, The twi-light shad-ows grey:
His Wis - dom beams on high, a star Is from its radiance wro't:

If God so clothe the ten - der flower, Now grow-ing, soon to die;
The brightwinged warblers of the grove, The for-est's sol - emn prayer,
The Star of Beth - le - hem ap - pears, To light the dark-en'd way

May we not trust our Fa-ther's power? Will he not hear our cry?
All whis - per of our Fa-ther's love, His ten - der, watch-ful care.
Of mil-lions, once in grief and tears, To im - mor - tal - i - ty.

Sweet 'Tis to Sing.

Words from DIAPASON by permission.

D. P. H.

1. Sweet 'tis to sing of Thee, Sa - vior and friend; Of thy great
love so free, Sa - vior and friend! Oh, for a heart to praise
Through all our earth - ly days, Thy won-drous works and ways,
Sa - vior and friend!

2. When thou wert here below,
   Savior and friend!
  Thou didst our sorrows know,
   Savior and friend!
  Grant to each heart to feel
  That thou hast power to heal;
  And, oh, thyself reveal,
   Savior and friend!

3. Tender and patient thou,
   Savior and friend!
  To thy dear love we bow,
   Savior and friend!
  Oh, in thy spirit pure,
  May we our ills endure,
  Trusting thy promise sure,
   Savior and friend!

4. By thy redeeming grace,
   Savior and friend!
  We hope to see thy face,
   Savior and friend!
  Then will we joyful praise,
  Throughout eternal days,
  Thy wondrous works and ways,
   Savior and friend!

Words by Dr. Blackall, # The Union Greeting. a. c. p.

1. Hith-er we come, as a Un-ion Band, To sing sweet songs of a
2. Greet-ing we give on this fes-tive night, A hap-py lay of the

bet-ter land, The land of peace and love; Where Je-sus reigns as a
heart's de-light, Good will on ev-ery hand; Bright eyes are beam-ing a-

King a-lone, And   all His chil-dren fond-ly own Their
mid the throng, And   young hearts glow as they sing the song Of

CHORUS.

Fa-ther, God a-bove. Oh! mer-ri-ly, mer-ri-ly,
this   our Un-ion Band. Oh! &c.

joy - ous and free, Sing we the song of the true;

Cheer - i - ly, cheer - i - ly, hap - py are we, Warm is our wel-come to

BOYS.   GIRLS.   TOGETHER.

you.   Wel - come! wel - come! Warm is our wel-come to you!

3. Gems have we brought to delight the soul,
   And flowers whose fragrance shall e'er be whole,
   That cheer life's way along;
   Then give your hearts and extend your hands,
   And let us bind you in silken bands,
   The bands of love and song.

Chorus—Oh! joyously, joyously sound we the strain,
   For 'tis the song of the true;
   Cheerily, cheerily give we again,
   Welcome, thrice welcome to you.
   Welcome! welcome!
   Welcome, thrice welcome to you.

"Feed My Lambs."

For Sunday School Concerts or Exhibitions. To be sung by eleven little girls, standing in the form of a crescent, each with a card or shield hung upon the breast, having upon it one letter of the Scripture motto which is the theme of the whole song. At first the letters are all reversed and unseen. But each singer turns her card and reveals her letter as she sings. All may sing the first six lines of the poetry if desired.

J. H. EDWARDS.

ANDANTINO

1. Christ you know, loved lit - tle children, When he lived on earth be - low,
3. In good works none should be backward, As you'll will-ing - ly a - gree,

And He gave to His dis - ci - ples A com-mand all ought to know.
So I've come my aid to ren - der, And have brought the let-ter E.

2. We have come this { day / night } to spell it Hap - py glad-some children we;
4. Still an - oth - er E is need-ed This command of Christ to spell;

I the let - ter F con - tri-bute, Here it is, as you may see.
Here it is, the need-ed let - ter, Can't you see it ver - y well?

5. Fourth among the list of letters
   Stands the one you ask of me;
   So I think 'twill not surprise you,
   When I show the letter D.

6. All my little friends above me
   Stepped from up the alphabet;
   I go half way down the column,
   And the letter M I get.

7. Farther down than all the others
   To the last but one I'll go;
   And the letter Y will furnish,
   Which completes two words, you know.

8. Next the letter L is wanted
   In the work we have to do;
   It begins the name Christ taught us,
   Here I turn it round to you.

9. Before all the other letters
   Is the one I bring you now;
   It is A, and lambs without it
   Cant be spelled, as you'll allow.

10. Once before upon the platform
    Has my letter been { to night / in sight }
    But another M is needed,
    So I'll turn it to the light.

11. Since my little friend above me
    In the line has called out A,
    'Tis but just a B to furnish,
    So I've brought it up this way.

12. Last of all in this procession,
    With the letter S I stand,
    Which, you know, completes the spelling
    Of our Saviour's blest command.

Chant (to be sung when the motto is complete.)

1. "Feed my lambs," 'twas Je-sus said it; "Feed my lambs" you read it here
*2. Jesus, gentle Shepherd, bear us, Bless these little lambs of thine;

That ye heed it and obey it, Let it in your lives ap-pear.
From all sin and danger keep us, Save us by Thy power di-ine. A - men.

* The 2d verse of the chant may be repeated by one little girl, all singing the "Amen;" or it may be chanted like the first verse.

"Children for the Union." F. B. Rice.

1. We are
2. We will

one and all for Un - ion, North and South, and East and West; All the
love our land for - ev - er, Dear-est land be-neath the sun; Foe-men's

States in loved com-mu - nion, Heart and hand with free - dom blest.
steel shall not dis - sev - er Youth-ful hearts that now are one.

CHORUS

Then join in the joy - ful hur - rah, Hur -

rah for the land of the free; For the Un - ion and peace, for

free - dom and law, Hur - rah for the land of the free.

3. We are all a band of brothers,
   All the States are sisters too,
   And in time there will be others
   That shall happy vows renew.
   CHORUS. Then join, &c.

4. Union now, and Union ever!
   True hearts now for Union all!;
   We will keep it safe and never
   Shall our glorious Union fall!
   CHORUS. Then join, &c.

The Glorious Light.

James R. Murray.

JOYFULLY.

1. A glo - rious light has burst a - round us, Joy - ful day!
2. We'll sing to God a ho - ly cho - rus, Joy - ful day!
3. The young and old come forth to hear us, Joy - ful day!

joy - ful day! We see the chains that would have bound us,
joy - ful day! Truth shines in ra - diant bright-ness o'er us,
joy - ful day! And isles a - cross the o - cean hear us,

Joy - ful day! joy - ful day! The spark - ling wine we
Joy - ful day! joy - ful day! A firm and daunt - less
Joy - ful day! joy - ful day! We'll spread the truth where

ne'er will crave, To touch, to taste is to en - slave; We
host we stand; Ye mil - lions join our glo - rious band, And
man is found, Bear it to earth's re - mot - est bound, 'Till

drink the foun - tain's crys - tal wave, Joy - ful, joy - ful
plen - ty then shall bless our land, Joy - ful, joy - ful
ev - ery wind shall catch the sound, Joy - ful, joy - ful

CHORUS.

day! Hur - rah! A glo - rious light has burst a - round us, Joy - ful

day! joy - ful day! We see the chains that would have

bound us, Joy - ful day! joy - ful day!

Was it right?

R. R. H.

MODERATO.

1. If the boys and girls will list-en,
2. All their books so new and pret-ty,
3. Then a-long the street came, sing-ing,

I will tell them in my song, Of a
Lay up-on the dust-y ground; They were
Such a mer-ry lit-tle lad; But his

sad thing that I no-ticed, As to school I came a-long;
torn, and soiled and tum-bled, As their own-ers pushed a-round;
song soon ceased its ring-ing; And his hap-py face was sad,

'Twas a fight! 'twas a fight! 'twas a fight! 'Twas be-
In such plight, in such plight, in such plight; That they
At the sight, at the sight, at the sight; And he

100

tween two lit-tle chil-dren, Who had fall-en out in play;
nev-er will be de-cent To be used in school a-gain;
part-ed them so gen-tly, And he begged them so, to cease,

And, a-las! they beat each oth-er In a
But the boys had both for-got-ten All a-
That they twined their arms to-geth-er, And all

rude and an-gry way! Was it right? was it right? was it right?
bout their les-sons then. Was it right? was it right? was it right?
went to school in peace. That was right! that was right! that was right!

Be Kind and True.—ROUND.

1 2 3

Be kind and true In all that you may do, Keep this in view.

Let it Pass.

C. H. Greene.

ALLEGRETTO.

1. Be not swift to take of-fense; Let it pass, let it pass, let it pass.
2. Ech- o not an an-gry word; Let it pass, let it pass, let it pass.
3. If for good you've taken ill; Let it pass, let it pass, let it pass.

An - ger is a foe to sense, Let it pass, let it pass, let it pass.
Think how oft-en you have err'd; Let it pass, let it pass, let it pass.
Oh! be kind and gen-tle still; Let it pass, let it pass, let it pass.

Brood not dark-ly o'er a wrong, Which will dis - ap pear ere long,
Since our joys must pass a - way, Like the dewdrops on the spray,
Time at last makes all things straight, Let us not re-sent, but wait,

Rath-er sing this cheer-y song—Let it pass, let it pass, let it pass.
Wherefore should our sorrows stay ? Let it pass, let it pass, let it pass.
And our triumph shall be great: Let it pass, let it pass, let it pass.

Willie and the Angels.

GENTLY

1. Wil - lie laid him down to sleep, When his eve - ning pray'r was said,
2. He had spok - en has - ty words, When his lit - tle sis - ter Sue
3. Now he sees her in his mind, With her blue eyes fill'd with tears—
4. Now he breathes his pray'r a - gain, Ask - ing par - don— seek - ing grace,
5. And a - gain an - oth - er eve, Hov - er'd an - gels round his bed;

And the gloom - y shad - ows creep Clos - er round his lit - tle bed.
Ask'd in voice so like a bird's, "Wil - lie, please to tie my shoe,"
How could he have been un - kind? How, so dread - ed i - die sneers?
When he whis - per'd his "a - men," There were an - gels In the place ;
He had not made Su - sie grieve, Nor a naugh - ty word had said,

In his heart, a shad - ow lay, He had not been good all day.
He would not have turn'd a - way, But the boys were there at play.
Oh! he wish - es Su - sie knew, How he longs to tie that shoe.
And the shad - ows fled a - way, From the couch where Wil - lie lay,
And they whis-per'd, "Let us stay, Wil - lie has been good all day."

In his heart, a shad - ow lay, He had not been good all day.
He would not have turn'd a - way, But the boys were there at play.
Oh! he wish - es Su - sie knew How he longs to tie that shoe.
And the shad - ows fled a - way, From the couch where Wil - lie lay.
And they whis - per'd "Let us stay, Wil - lie has been good all day."

Have you sold your matches Tom?

G. F. R.

1. Are all your match-es sold yet, Tom?
2. We'll call the sun our fa - ther, Tom,
3. But Oh, there's One a - bove him, Tom,
4. We'll tell Him all our sor - rows, Tom,

Are all your match - es done? Then let us to the
We'll call the sun our mother; We'll call each pleas - ant
Who loves us more than he; Who made the great bright
We'll tell Him all our care; We'll tell Him where we

o - pen square, And warm us in the sun; To
lit - tle beam, A sis - ter or a brother: He
sun to shine, With beams so warm and free; He
sleep at night, We'll tell Him how we fare: And

warm us in the sweet bright sun, To feel his kin - dling
thinks no shame to kiss us, Tom, Al - though we rag - ged
is our re - al fa - ther, Tom, Al - though while here be-
then, Oh then to cheer us, Tom, He'll send his sun to

glow; For his kind looks are the on-ly looks Of
go; For his kind looks are the on-ly looks Of
low; The sun's kind looks are the on-ly looks Of
glow; For his kind looks are the on-ly looks Of

friend-ship that we know. O Tom, don't you cry, Al-

CHORUS.

though the cold winds blow; For the sun is shin-ing

bright and warm, In the great square down be-low.

The Blind boy.

[Arranged from "*Rosa Lee*," by permission of D. P. FAULDS. LOUISVILLE.]

MODERATO

1. One day I saw a lit-tle child Low sit-ting in an
2. To him there came no joy-ous sight; He could not mark the
3. He said "When morn-ing bright-ly breaks I gai-ly sing a
4. I turned a-side, when this I heard, And went up-on my

o-pen door; And such a face, So meek and mild, I
speed-ing ball; He could not see the soar-ing kite, Nor
pleas-ant song; My listening soul to mu-sic wakes, And
on-ward way, But oft the lit-tle suff'rer's word Came

thought I ne'er had seen be-fore. And sport-ing on the
watch the glistening wa-ter-fall, He could not trace the
then I hear it all day long. I lis-ten to the
to me in the bus-y day. And when I saw how

sun-ny green, So bright the day, the scene so fair, Gay
Rob-in's flight, A-cross the sky, so free and glad. For
sigh-ing breeze, I heark-en to the wa-ter-fall. And
dis-con-tent Had cloud-ed many a youth-ful brow, To

groups of boys and girls were seen. I asked him why he
him was one un - end-ing night. I asked him if he
up a - mong the whisp'ring trees, I hear the mer - ry
whom all bless-ings had been sent, I thought that I would

ling - er'd there? His soft brown hair he put a - side, And
was not sad? A smile shone o'er his gen - tle face, Of
bird - lings call, And so all day in joys of sound, My
tell them how Peace, like a gen - tle riv - er's flow, Had

lis - ten'd to my ques-tions kind; And then,—in ac - cents
joy I had not thought to find, And then he said with
soul a so - lace seeks to find: And Oh! while mu - sic
made the blind boys heart re-signed, And taught him in his

low, re - plied, "I can - not play, for I am blind!"
mod - est grace, "Oh no I'm hap - py, though I'm blind!"
swells a - round, I am so hap - py, though I'm blind!"
bit - ter woe, To seem so blest, though he was blind!

Morning Prayer.

By permission S. T. GORDON.   D. P. H.

1. Heaven - ly Fa - ther! throned a - bove us,

Hear our prayer, to thee ad-dressed; Let me now, and

through life's jour - ney, In thy love se - cure - ly rest.

2. Thy unfailing love and mercy,
    Are to me a shining light ;
    And thy word. the dew of heaven,
    Glistening ever pure and bright.

3. Thou dost kindly clothe and feed me,
    To my wants thou givest heed ;
    Grant me, Lord, thy richest blessing,
    And sustain me in my need.

4. Let thy spirit guide me ever,
    Teach me to obedient be ;
    Bring me to thy heavenly kingdom,
    There to dwell eternally.

1. We're the lambs of the flock, and no dan - ger we fear,
2. O, the pas-tures are green, and the flow'rs bloom a - round,
3. O, that all the dear lambs had a heart to re - ply,

When the voice and the call of the Shep - herd we hear.
By the side of still wa - ters He lets us lie down.
When the great Shep - herd calls from his man - sions on high.

CHORUS

Then we fol-low, then we fol-low, then we fol-low, fol-low, fol-low,
Then we fol-low, then we fol-low, then we fol-low, fol-low, fol-low,
We will fol-low, we will fol-low, we will fol-low, fol-low, fol-low,

fol - low, In the steps of the flock, when the Shep-herd we hear.
fol - low, Then we fol - low His call, where the flow'rs bloom a-round.
fol - low, We will fol low the call to His fold in the sky.

The Happy Morn We Hail Again.

By permission of S. T. Gordon.

CHEERFUL, BUT NOT TOO FAST.

1. The hap - py morn we hail a - gain, When heav'n seems smil-ing o'er us;
2. And with the hum-ble shep-herd throng, A-round his cra-dle man - ger,

And from the sky in joy - ful strain Breaks forth the an-gel cho - rus.
We gath - er now with pray'r and song To greet the in - fant stran-ger.

Peace on earth, good will to men; Glo - ry in the high - est.

3. We bring no gems, nor rich perfume,
Nor wisdom's years before him;
But come in childhood's early bloom,
In childhood's praise adore him.
Peace on earth, &c.

4. For thou who wert thyself a child,
In more than infant meekness;
Wilt never in thy mercy mild,
Despise our childhood's weakness.
Peace on earth, &c.

5. O, send thy Sp'rit, us to bless,
That in thy footsteps holy;
Our feet may turn to righteousness
From paths of sin and folly.
Peace on earth. &c.

6. Then, led by thee, our souls shall rise,
Where thou hast gone before us;
And bless thee ever in the skies,
That earth has heard the chorus.
Peace on earth, &c.

G. F. R.

1. Be-yond the dark riv-er of death— Be-yond where its wa-ters are swell-ing, The

home of my spir-it is wait-ing for me, The land where the ran-som'd are dwelling.

CHORUS.

No night in that beautiful home! No shade on its glo-ry is seen; The

wonderful riv-er of wa-ter of life Flows soft thro' the meadows of green.

2 No grief in that beautiful home!
 No sorrow can enter its portals!
But glad are the voices that join in its
 song,
 The song of the shining immortals.
  Chorus.

3 No tears in that beautiful home,
 No sin from our Savior to sever!
The King in his beauty our eyes shall
 behold,
 And join in his praises forever!
  Chorus.

Will You Come?

Words by MRS. PRAY.   (BY REQUEST.)   Arranged by D. P. H.

1. Come, sin - ner, to the Sa - vior—come; Will you come?
   He calls you from the heav'nly throne; Will you come?

Will you come? To Je - sus come, for - sak - ing sin, And

strive e - ter - nal life to win. Will you come! Will you come?

2. Come willing, now, from sin to part;
   Will you come? will you come?
   Obey his call—give him your heart;
   Will you come? will you come?
   In humble penitence, come prove
   His tender, yearning, boundless love.
   Will you come? will you come?

3. His love would all your sins forgive;
   Will you come? will you come?
   You now the blessing may receive;
   Will you come? will you come?
   With rapture, feel the love divine,
   While soft he whispers thou art mine.
   Will you come? will you come?

Cold Water.

By permission of S. T. GORDON.

D. P. H.

SPIRITLY.   SOLO OR DUET.

1. The flow-ers drink At the stream-let's brink, And the oak leaves drink the dew;

And the song-sters sing of the spark-ling spring, As they soar in the a-zure blue.

CHORUS

O, the wa-ter cold, With its wealth un-told, From the earth out-gushing free.

As it bubbles and sings in a thousand springs, Is the drink, the drink for me.

2. The sunlight sleeps where the rain-
    king keeps
His treasures uplaid in the sky;
Or it bids a bow in its beauty glow,
When the storm sprite passes by.
    O, the water cold, etc.

3. We'll point to the spring, as we 'join
    to sing,
And repeat and pledge again,
"All things we hate that intoxicate,"
To the burden of our strain.
    O, the water cold, etc.

Anniversary Day.

Words by Mrs. Pray.

D. P. H.

1. Passed an - oth - er chang - ing year, Its pleasure and its pain,
2. Some who but a year a - go U - nit - ed in our throng,
3. We are tend-ing to their home, Where years no changes bring;
4. Bless-ed Je - sus, make us wise To live on earth for heaven;

Hap - py chil-dren gath - er here, To sing their songs a - gain.
Now the joys of an-gels know, And min - gle in their song.
Loved ones wait for us to come With them their songs to sing.
None need fear how soon he dies, Whose sins thou hast for - given.

Sing to - day with sa-cred joy, Think of loved ones passed a - way;

Songs of joy the hours em-ploy On An - ni - ver - sary Day.

Words by Mrs Pray.   D. P. H.

1. Precious time! why is it given, But to fit my soul for heaven?
2. If, to save my soul in heaven, Je-sus' life and death were given,

How shall I its hours em-ploy To se-cure my high-est joy?
Now to him my-self I give; He will show me how to live,

If on earth I may pre-pare All the bliss of heaven to share,
And a treas-ure to se-cure That shall ev-er more en-dure.

On-ly this will I pur-sue—Here and now my work to do.

3. Here he lived his life for me,
His disciple would I be;
For the glory of his cross
Learn to count all else but loss;
He will stoop from heaven above,
Teach me how like him to love;
Only this will I pursue—etc.

4. Work for Jesus, here and now;
Greater joy can heaven bestow?
There no suffering, no betrayed,
Sick nor sad ones need my aid.
Here my pitying love may bless
Every form of wretchedness;
Only this will I pursue—etc.

Because He Loved Me So.

G. F. R.

MODERATO.

1. I love to hear the sto - ry Which an - gel voi - ces tell,
2. I'm glad my bless - ed Sav-ior Was once a child like me,
3. To sing his love and mer - cy, My sweet-est songs I'll raise,

How once the King of Glo - ry Came down on earth to dwell:
To show how pure and ho - ly His lit - tle ones might be:
And though I can - not see him, I know he hears my praise!

I am both weak and sin - ful, But this I sure - ly know,
And if I try to fol - low His foot - steps here be - low,
For he has kind - ly prom - ised That I shall sure - ly go,

The Lord came down to save me, Be - cause he loved me so.
He nev - er will for - get me, Be - cause he loves me so.
To sing a - mong his an - gels, Be - cause he loves me so.

O, Jesus! Light of All Below!

"I am the light of the world."—John, 8: 12.

Old Latin Hymn.   Arr. from *Laudi Spirituali*, by Rev. J. T. Duryea, D. D.

1. O, Je-sus! light of all be-low! Thou fount of life and fire!
2. May ev-'ry heart con-fess thy name, And ev-er thee a-dore;

Sur-pass-ing all the joys we know—All that we can de-sire;
And seek-ing thee, it-self in-flame To seek thee more and more.

When once thou vis-it-est the heart, Then truth be-gins to shine;
Thee may our tongues for-ev-er bless; Thee may we love a-lone;

Then earth-ly van-i-ties de-part, Then kin-dles love di-vine.
And ev-er in our lives ex-press The im-age of thy own.

ANDANTINO

1. Young Bol - ter Brook is   a   beau - ti - ful brook While the
2. You launch your boat on   the   per - i - lous tide, With a
3. There   is   a   boy that   I   hap - pen to know, Ver - y
4. He   comes to school at   the   nine o - 'clock bell, And   he

A - pril rains do pour; But   when the rain stops, His courage all drops
will the oars to try, The   foam and the spray, All van-ish   a - way
much like Bol-ter Brook, As   read-y   a   lad As need to   be had,
real - ly means to try; But   when the bell stops, His courage all drops,

And Bol - ter is   no more. Oh   the long pull, and the strong pull, No
And leave you high and dry. Oh   the long pull, &c.
But will not mind his book. Oh   the long pull, &c.
And leaves him high and dry. Oh   the long pull, &c.

mat - ter what's the weath - er,   Is   the   glo - rious way   to crown the day and we'll all march on to - geth - er, Is the

glo-rious way to crown the day and we'll all march on to - geth - er.

Good Night.

G. F. R.

ANDANTINO.

1. Come, let us sing a pleas-ant song, As to our homes we go a-long;
2. We'll seek in peace each qui-et home, For now the evening shades have come;
3. Yes, dear com-pan-ions, fare you well, A-gain our part-ing num-bers swell;

With cheer-ful tones and spir-its light, We'll sing a-gain our glad good-night.
With cheer-ful tones and spir-its light, We'll sing a-gain our glad good-night.
With cheer-ful tones and spir-its light, We sing once more our glad good-night.

Good Night.—CONCLUDED.

REPEAT. *pp*

Good night, good night, good night, good night.

Good night, good night,

Excursion Song.

To be sung while approaching the grove.

B. R. H.

ALLEGRETTO

1. Ho! ho! ho! Out to the beau - ti - ful groves we go;
2. Sing! sing! sing! Heav - en shall smile at the praise we bring;
3. Play! play! play! Run, oh, ye hap - py ones while ye may;

This is our hol - i - day now, you know; Sweet shall our mel - o - dies
For - est and mead-ow with mu - sic ring, Ech - o the cad - en - ces
Roam thro' the for - ests at will to - day, Pour-ing your shouts and your

float and flow, Out on the balm - y air:
grace - fully fling, Out on the balm - y air:
laugh - ter gay, Out on the balm - y air:

Bear them, ye breez-es that gen - tly blow, Scat - ter them ev - 'ry-
Bear them a - loft on her sil - v'ry wing, Scat - ter them ev - 'ry-
Syl - vi - a beck-ons, oh speed a - way, Scat - ter them ev - 'ry-
where; Bear them, ye breez - es that gen - tly blow,
where; Bear them a - loft on her sil - v'ry wing,
where; Syl - vi - a beck - ons, oh speed a - way,
gen - tly blow, gen - tly blow; Bear them, ye breez - es that
sil - v'ry wing, sil - v'ry wing; Bear them a - loft on her
speed a - way, speed a - way; Syl - vi - a beck - ons, oh
gen - tly blow, Scat - ter them ev - 'ry - where.
sil - v'ry wing, Scat - ter them ev - 'ry - where.
speed a - way, Scat - ter them ev - 'ry - where.

Fitly Spoken Words.

Words by Mrs. Pray.   D. P. H.

1. Fit - ly spo-ken words, O, use them   For the Mas-ter, Prince of Peace;
2. Not with words of sol - emn sad-ness Seek, where mirth is, to do good,
3. To the suff'rer, faint and dy-ing, Whis-per while he yet can hear,

He will teach you   how to choose them, Bidding strife and   discord cease,
While the hours are wing'd with gladness, Meet the mer - ry   in their mood.
Words in which he   may, re - ly - ing, Pass from earth with - out a fear.

Learn the precious   min - is - tra-tion, How with words to   give re - lief.
Nev - er dare sup - press, no, nev - er, Words the err - ing   to re-strain,
When you reach the heav'n-ly por - tals With the souls your words have won,

In the hour of   trib - u - la - tion, To the heart, surcharged with grief.
Tho' dis-tress-ing   the en-deav - or, Speak, you may your brother gain.
Welcom'd to the   joys im-mor - tal, You shall hear the words, "well done."

Come to Jesus.   123

D. P. H.

1. Come to Je-sus, ye who wan-der Far from hope, and peace and rest;

Scorn'd, neg-lect-ed, and for-sak-en, Sor-row-ful and sore dis-tress'd.

Come to Je-sus! come to Je-sus! Ye by sin and fear op-press'd.

2. Come to Jesus! he hath loved you
   With an everlasting love,
   And his heart of tenderest pity
   Needs no sacrifice to move.
      Come to Jesus! come to Jesus!
   And his free salvation prove.

3. Come, oh! come! the Master waiteth,
   Come, the longing bride doth say;
   Come, he tarries while we linger,
   He hath borne our sins away.
      Come to Jesus! come to Jesus!
   Come, the spirit cries, to-day.

Only Believe.

Words by MRS. PRAY.

D. P. H.

1. On - ly be - lieve, are the words of the Sav - ior;

Sad one, oh! has - ten, his words to re - ceive:

Faith in his word is the key to his fa - vor;

Oh! be not faith - less, but on - ly be - lieve.

CHORUS.

Je - sus in - vites you, now heed - ing his voice,

Sor - row no long - er— re - joice, oh! re - joice.

2 Fettered by sin, in repentance you languish,
   Yearning the life of the righteous to live:
   Jesus has witnessed your heart-rending anguish;
   Cease from your struggling, and only believe.
   Jesus invites you, etc.

3 Trusting his love, will your sins be forgiven;
   Freely he suffered, your soul to retrieve;
   Died to redeem you, and save you in heaven—
   Yours is salvation, now—only believe.
   Jesus invites you, etc.

The Word.

G. F. R.

MODERATO

Oh may we Thank-ful be, For the won-drous fa - vor,
Of Thy word, Blessed Lord, Mak-er, King and (omit) Sav - ior.

A Father in Heaven.

(Originally written for the Children of the Brooklyn Orphan Asylum.)

Words by MRS. PRAY.

D. P. H.

1. The Fa - ther of An - gels in heav - en, is ours,

He loves lit - tle chil - dren to bless;

We love Him who bright-ens our world with sweet flow'rs,

In prayer we will seek his ca - ress.

CHORUS.

A Fa - ther in heaven! A Fa - ther in heaven! We'll

pray to our Fa - ther, Our Fa - ther in heaven.

2 Our Father will hear us in pitying love,
   His promise forever is true;
   His eye, ever watchful, will always approve
   The good we are striving to do.
   A father in heaven, etc.

3 Through life and in death this dear Father will love,
   And keep us with tenderest care;
   And cheer us with hope of the mansions above,
   Which He has said He will prepare.
   A father in heaven, etc.

Be You to Others.—ROUND, IN THREE PARTS.

Be you to oth - ers kind and true; And al - ways

un to oth - ers do, As you'd have oth - ers do to you.

Invocation to the Spirit.

Old Latin Hymn.   Arr. from *Laudi Spirituali*, by Rev. J. T. Duryea, D. D.

1. Come, heav'n-ly Spir - it, come! Come un - to me!

Kind Fa - ther of the poor, Come un - to me,

Giv - er and bless - ed Gift, En - ter my low - ly door!

Be guest with - in my heart, Nor ev - er hence de - part;

Come, heav'n-ly Spir - it, come! Come un - to me.

2 Unveil thy glorious self
    Come unto me!
  To me, O Holy One,
    Come unto me!
  That thou into my heart
    May shine thyself alone!
  Saved from earth's vanities,
  To thee I long to rise.
    Come, heavenly Spirit, come!
    Come unto me.

3 Renew me, Holy One!
    Come unto me!
  O purge me in thy fire;
    Come unto me!
  Refine me, heavenly flame,
    Consume each low desire;
  Prepare me as a sacrifice,
  Well pleasing in thine eyes.
    Come, heavenly Spirit, come!
    Come unto me.

Pilgrim.

1. Pilgrim, burden'd with thy sin, Haste to Zi-on's gate to-day;
2. Knock! for Mer-cy lends an ear; Weep! she marks the sin-ner's sigh;

There, till Mer-cy let thee in, Knock and weep, and watch and pray
Watch! till heav'nly light ap-pear; Pray! she hears the mourner's cry.

3 Mourning pilgrim! what for thee
  In this world can now remain?
  Seek that world from which shall flee
  Sorrow, shame, and tears and pain.

4 Sorrow shall forever fly:
  Shame shall never enter there;
  Tears be wiped from every eye;
  Pain in endless bliss expire.

The Pilgrims' Planting.

Words by Rev. E. Whitaker.

D. P. H.

1. O - ver the sea to un - known shores, Ex - iles of
2. Here shall that Cross for - ev - er stand, Sym - bol of

faith the Pil - grims came, Free - dom they sought, not
life to dy - ing souls, Firm as a rock mid

gold - en ores, God's book their law; their trust, his name.
shift - ing sands, Where in his wrath the O - cean rolls.

Sigh - ing, they left their Fa - ther - land, Trac - ing the
Vi - tal and fair a - bides that tree, Throw - ing its

flight of Lib - er - ty, Here on this spot that
arms to ev - 'ry wind, Un - der its shade for

faith - ful band Plant - ed the Cross and Free - dom's Tree.
aye shall be Rest and de - light for all man - kind.

Content. D. P. H.

1. Why should my fleet-ing mo-ments be In vain re-pin-ings spent?
2. The bless-ings that may crown my lot, I would re-gard as lent;

Oh, teach me, Lord, whate'er my fate, To be therewith con - tent.
En - joy them with a grate-ful heart, And be with them con - tent.

3 And though thy providence remove
  The good thy hand hath sent,
Still would I raise a trusting eye,
  And banish discontent.

4 Of gloomy thoughts and murm'ring
  O Lord! I would repent; [words,
And henceforth sing the pilgrim's song
  Of faith and sweet content.

The Spires of the Golden City.

ANDANTINO.   G. F. R.

1. Dawn on our vi - sion, O,
   Dwell - ing in bright - ness out -
2. Far in the dis - tance, that
   Seen by the pil - grim, when

cit - y of light, Throned in the land where "there shall be no night;"
shin - ing the sun, Glad in the smile of the in - fin - ite One:
glo - ry we hail, Seen from the depths of hu - mil - i - ty's vale;
wea - ry and faint, Pour - ing his soul out to heav - en in plaint:

Dawn on our vi - sion, when tempt - ed to stray, Show - ing the
O - ver each toil - some as - cent he must climb, O - ver the

end of the nar - row - er way; Show - ing their home who that
bil - lows of death and of time, Gleam - eth the pearl of the

132

path-way have trod, Home of the an - gels and Cit - y - of God,
por - tals of rest, Gold-en - ly gleam-eth the spires of the blest.

CHORUS

Beau - ti - ful cit - y! thy spires from a - far Gleam on the

vi - sion as Beth - le - hem's star; Un - der the scep - ter of

grace, or the rod, Press we to gain thee, O, Cit - y of God.

3. When the last tears of this earth-life are shed,
When its last sun in the even is red,
Who may to ear of the living unfold,
What those rapt eyes of the dying behold?
See they not visions of light and of love?
Lost ones of earth that are angels above?
See they not over the portals of rest,
Goldenly gleaming, the spires of the blest?
    Beautiful city! &c.

Come.

G. F. R.

GENTLY

1. On the ear of child-hood fall - ing, List, and you shall hear
2. When the way grows dark and drea - ry, Hark, the call shall be,
3. When Thou from our sin hast freed us, By Thy wondrous love,

Je - sus gen - tly call - ing, call - ing Lit - tle chil - dren near.
"Hea - vy - la - den, lone and wea - ry, Come thou un - to me!"
Bless - ed Sa - vior, guide and lead us To thy courts a - bove.

Lov - ing - ly our Lord be - holds them! Kind - ly bids them come!
Sweet - ly Je - sus call - eth, call - eth, "Come thou to my breast;"
So when on our ear is fall - ing, "Come! ye bless-ed, come!"

In His arms of love He folds them; 'Tis the chil-dren's home.
Peace-ful - ly the prom - ise fall - eth, "I will give you rest!"
We may go when Thou art call - ing, Home, for - ev - er home!

Father and Savior.

G. F. R.

And God said "Suffer little children to come unto me, and forbid them not, for of such is the kingdom of heaven."

MODERATO

1. My Fa - ther and my Sav - ior, From fields of heavenly light,
2. His eye is ev - er o'er me, Nor slum-bers night nor day;
3. Oh, may I tru - ly love Him, And flee from all that's bad,

Is look - ing down up - on me To see if I do right:
He sees each look and ac - tion, And hears each word I say:
And then He'll smile up - on me And make my spir - it glad:

He stands in ho - ly gran - deur A - mid the an - gel throng;
I know my heart is way - ward, Its wrong has ne'er been told:
He'll keep me from all e - vil, And guide me on my way,

Oh! may I weep with sad - ness When - ev - er I do wrong.
Yet oh, He ev - er loves me And calls me to His fold.
And I will kneel and ask Him To teach me how to pray.

Take up thy Cross.

MODERATO.

1. On - ward, still on - ward, the path - way is straight,
2. Thorn-pierced and wea - ry, no pause for a day,
3. Know thou art near - ing the riv - er, per - chance

Nar - row the way to the beau - ti - ful gate;
Price - less the blood that is mark - ing the way;
On - ly the march of a day in ad - vance;

Cheer'd by thy fel - lows, or press - ing a - lone,
What tho' the voice of af - fec - tion should urge
O - ver the flood, lo! the shin - ing ones wait,

Fal - ter not, faint not, the goal is thine own;
Sun - ni - er paths, thou wouldst find them di - verge;
With thee to en - ter the beau - ti - ful gate;

Meet ev - 'ry tri - al of grief and of loss,
Cares and af - flic - tion may buf - fet and toss,
Earth is a cru - ci - ble, purg - ing thy dross

Pa - tient - ly, pray'r - ful - ly— take up thy cross.
Greet them as ' heav - en - sent— take up thy cross.
Heav - en has crown for thee— take up thy cross.

O give Thanks unto the Lord.—Chant.   G. F. R.

1. O give thanks unto the Lord, for ... | he is good: | For his | mer-cy is for- | ev - er
3. O give thanks unto the .... | Lord of lords: | For his | mer-cy is for- | ov - er
5. To him that by wisdom.................. | made the heav'ns: | For his | mer-cy is for- | ev - er.
7. To him that................................... | made great lights: | For his | mer-cy is for- | ev - er.
9. The moon and stars to.................... | rule by night: | For his | mer-cy is for- | ev - er.

2. O give thanks unto the................. | God of gods: | For his | mer-cy is for- | ev - er.
4. To him who alone........................... | doeth great wonders: | For his | mer-cy is for- | ev - er.
6. To him that stretched out the earth a- | bove the waters: | For his | mer-cy is for- | ev - er.
8. The sun to..................................... | rule by day: | For his | mer-cy is for- | ov - er.
10. O give thanks unto the Lord, for... | he is good: | For his | mer-cy is for- | ov - er.

138   **To Thee, O God.**   J. R. M.

1. To Thee, O God, we of - fer Our joy - ful songs of praise;
2. Guard Thou the young, we pray thee, From sin and er - ror's ways;

To Thee, the boun-teous Giv - er, And Guard-ian of our days:
Show them the path of du - ty, And guide thro' fu - ture days:

A - gain we meet to thank Thee, To raise our ev'n - ing pray'r;
May youth and age so serve Thee, Thou God of watch-ful love;

Our hearts are fill'd with glad - ness For Thy most ten - der care.
That they, when life is end - ed, May dwell with Thee a - bove.

"Thou Knowest that I Love Thee." f.w.r. 139

1. Thou know-est how each word of Thine Hath pierc'd the soul be-fore Thee;
2. Thou know-est earth was count-ed dross, As heav-en's hopes drew near-er;
3. Thou know-est what I can-not speak, For grief my soul is fill-ing;

Thou know-est how this heart of mine Would in the dust a - dore Thee;
That I have meek - ly borne the cross, Thy frail re-pent-ant hear - er:
Thou know-est that the flesh is weak, The spir-it tru - ly will-ing;

That it has bow'd be-neath Thy rod, And owns no tho't a - bove Thee;
Thou know-est that my home—my all, Was not en-thron'd a-bove Thee:
That I wo'd tread where Thou hast trod, Tho' ru - in frown a-bove Thee;

Thou know-est all things, O, my God, Thou know-est that I love Thee.
That I have heark-en'd to Thy call; Thou know-est that I love Thee.
Thou know-est all things, Sav-ior, God, Thou know-est that I love Thee.

From the Shining Morning Land.

G. F. R.

MODERATO

1. From the shin - ing Morn-ing Land, To this hap - py child-hood band,
2. We have come our joy to tell, And our bliss - ful num-bers swell;
3. Chil-dren dear, O, watch and pray; Live for Je - sus ev -'ry day;

We have come to join your cho - ral notes of song;
While we wave our shin - ing ban - ners in the light.
You shall join us in the bright and beau - teous sky:

Loud ho - san - nas we will sing, While the star - ry man-sions ring,
But our words can nev - er show, All the joy our bo-soms know,
We will lead you up the way, Up the shin-ing, heav'nly way.

And the cho - rus of your swell - ing strains pro - long.
While we gaze in rap - ture on this hap - py sight.
To the man - sions of the bless - ed ones on high;

REFRAIN.

We have come to join your lay, On this peace-ful Sab-bath day;

Come to join your hap-py, hap-py child-hood band; 'Tis the

same Re-deem-er's love, Thrills our rap-tured souls a-bove,

I will Seek My Father.

From Blumenthal, by F. W. R.

REVERENTIALLY.

1. When the morn is bright and fair, When sweet songsters charm the air,
2. In the sol - i - tude a - part, In the wil - der - ness or mart,
3. When the ev'n-ing sun is red, When each blos-som droops its head,

I will lift my heart in pray'r, I will seek my Fa - ther;
Oh! my sore - ly tempt - ed heart, I will seek my Fa - ther;
Kneel-ing low be - side my bed, I will seek my Fa - ther;

Lest my feet should go a - stray From His pure and per - fect way;
In the darkness as the day, He shall be my Guide and Stay;
That I slumber in His care, Shield-ed from each harm ful snare;

Lest I grieve Him as I may, I will seek my Fa - ther.
I will lean on Him al - way— I will seek my Fa - ther.
And for life or death pre - pare; I will seek my Fa - ther.

CONTENTS.

www.ingramcontent.com/pod-product-compliance
Lightning Source LLC
Chambersburg PA
CBHW021817190326
41518CB00007B/639